C0-CEU-350

CHICAGO PUBLIC LIBRARY
HAROLD WASHINGTON LIBRARY CENTER
R0021952694

Fixation
in Histochemistry

Fixation in Histochemistry

Edited by

P. J. Stoward

Senior Lecturer in Anatomy
University of Dundee
and Editor of The Histochemical Journal

Foreword by Professor A. G. E. Pearse

CHAPMAN AND HALL · LONDON
1973

11 New Fetter Lane, London EC4P 4EE
Reprinted, with additional foreword,
preface and index, from The Histochemical Journal
(published by Chapman and Hall),
Volume Four, Numbers Four, Five, and Six,
and Volume Five, Number One.

SBN 412 12050 X

Distributed in the U.S.A.
by Halsted Press, a Division
of John Wiley & Sons, Inc.,
New York

Contents

Foreword

by Professor A. G. Everson Pearse

The title of the Symposium, at which the papers which compromise this volume were presented, is of the utmost importance. It was not 'Fixation and Tissue Destruction' or 'Fixation and Loss of Tissue Components', but 'Fixation and *Tissue Preservation*'.

Historical fixatives, some still with us in the field of light microscopy after over 100 years, are not less guilty than the new historical 'fixatives' of ultrastructural practice some of which remove up to 60% of the material originally present in the specimen and convert the remainder into chemically inert products.

There must be few histochemists who do not appreciate the great need for improvements in the practice of fixation, but whence can we hopefully expect the arrival of such improvements?

Apart from out of the blue, or revolutionary, advances such as the advent of glutaraldehyde, there are perhaps three principal sources. The first is through theoretical, that is to say chemical and physical, considerations, involving practical action as a direct consequence. The second is through empirical experimental, that is to say trial and error, procedures and the third, through manœuvres designed to reduce fixation to a minimum. Our understanding of the processes involved in fixation is limited and our need for such understanding is increased by the needs of such disciplines as immunofluorescence and electron immunocytochemistry. Here the paramount need for preservation of a protein (antigen) conflicts with the absolute necessity to leave intact the antigenic determinant site. Similarly, in the case of enzyme cytochemistry, at all microscopic levels, the need to preserve a protein (enzyme) conflicts with the absolute necessity to leave intact the active site of the enzyme.

The present volume contains food for thought in relation to all the proposed sources of improved fixation, including the revolutionary ones. It should command the interest not only of histochemists but also of microscopists in many different sciences. I would hope that some of its readers might be stimulated by their reading to go out and do even better.

Department of Histochemistry,
Royal Postgraduate Medical School,
Hammersmith Hospital, London.

Preface

The chapters in this book, the first to be published on fixation, are essentially reprints of articles that have appeared in recent issues of *The Histochemical Journal.* With two exceptions, the articles are expanded versions of papers presented at a Symposium entitled 'Improvements in fixation and tissue preservation for histochemistry' organized by Professor J. R. Garrett on behalf of the Histochemistry and Cytochemistry Section of the Royal Microscopical Society at Oxford on 13 and 14 April 1972.

Most of us are aware of the problems of fixation but are reluctant to face them, hoping that others will tackle and solve them instead. We cannot afford to wait any longer. Equipment, especially electron microscopes and microanalytical instruments, and localization techniques are becoming more and more sophisticated, but their application is being held back by not knowing how to fix and preserve tissue specimens in order to obtain produceable results from which the true *in vivo* structure, composition and function can be inferred with certainty. There are two basic problems here. The first is to decide the criteria as to what constitutes 'good' fixation. Subjective assessments of morphology are not enough. Objective measurements are required, of the kind Drs Davies, Bone & Ryan describe in their articles.

The second problem is to unravel the effects fixation and tissue preservation procedures have on the chemical and physical properties of the various constituents of tissues. This is a complex subject, as Dr Jones clearly demonstrates in his chapter. Nevertheless, Dr Hopwood reveals, a considerable insight into such matters can be obtained with simple experimental procedures. However, for the time being, most biologists must content themselves with more pragmatic approaches such as are described by Drs Miller & Davies and Professors Avrameas & Garrett.

Fixation is often regarded as a necessary evil. That it can be turned to advantage is illustrated by Professor Ericsson's work on tissue culture cells. But at the end of the day, some biologists may still despair. They will sympathize with Professor Dawson's *cri de cœur:* 'Fixation: what should the pathologist do ?' Enough is written in this book to show that each of us can and should do *something*, and after that, do even more. If the hopes Professor Pearse expresses in his Foreword are fulfilled, hopes which I share, this book will have to be revised so as to give a balanced and up-to-date review of the whole state of the art; perhaps by that time, the art will have become a science.

P. J. S.

Dundee
January 1973

Contributors

Dr Stratis AVRAMEAS
Now: Unité d'Immunocytochimie, Département de Biologie Moléculaire, Institut Pasteur, Paris

Dr Q. BONE
Marine Biological Association of the U.K., Citadel Hill, Plymouth

Dr Ulf T. BRUNK
Department of Pathology, University of Uppsala

Dr D. F. DAVEY
Now: Department of Physiology, Monash University, Clayton, Victoria, Australia

Dr K. J. DAVIES
Department of Oral Pathology, The Dental School, King's College Hospital Medical School, London

Professor I. M. P. DAWSON
Now: Department of Pathology, University of Nottingham

Professor Jan L. E. ERICSSON
Department of Pathology at Sabbatsberg's Hospital, Karolinska Institutet Medical School, Stockholm

Professor J. R. GARRETT
Department of Oral Pathology, The Dental School, King's College Hospital Medical School, London

Dr David HOPWOOD
Now: Department of Pathology, University of Dundee

Dr Dennis JONES
N.V. Organon, Oss, Holland

Dr H. R. P. MILLER
Department of Veterinary Pathology, University of Glasgow

Dr K. P. RYAN
Marine Biological Association of the U.K., Citadel Hill, Plymouth

Reactions of aldehydes with unsaturated fatty acids during histological fixation★

DENNIS JONES

N.V. Organon, Oss, Holland

Synopsis. Formaldehyde reacts with unsaturated fatty acids in tissues during histological fixation. For example, the reaction of formaldehyde with oleic acid gives rise to compounds with the structures shown below. The structures of the major reaction products have been confirmed, but those of the minor products have not been conclusively demonstrated. Esterified unsaturated fatty acids react more slowly, but the reaction is otherwise similar. Acrolein has been found to react in a similar fashion, but the reaction is far more complex.

The occurrence of such compounds in tissues partly explains the loss of lipids during fixation, and raises some interesting possibilities regarding new fixatives and new histochemical reactions.

Major products

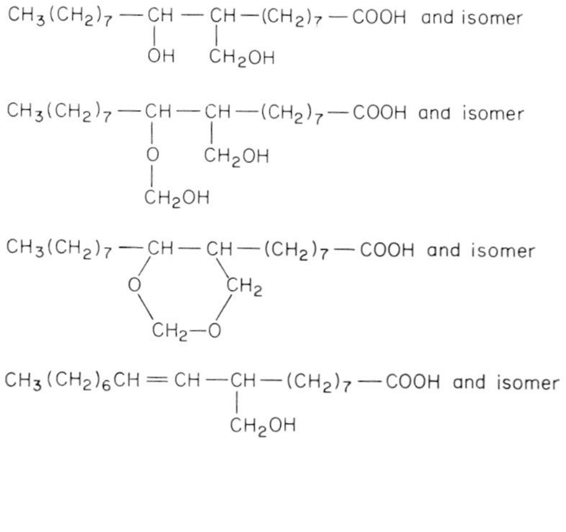

```
CH3(CH2)7—CH — CH—(CH2)7—COOH  and isomer
          |     |
          OH    CH2OH

CH3(CH2)7—CH—CH—(CH2)7—COOH  and isomer
          |   |
          O   CH2OH
          |
          CH2OH

CH3(CH2)7—CH—CH—(CH2)7—COOH  and isomer
         /     \
        O       CH2
         \     /
         CH2—O

CH3(CH2)6CH = CH—CH—(CH2)7—COOH  and isomer
                 |
                 CH2OH
```

★ The experimental work described in this paper was performed in the Departments of Pathology and Applied Biology, University of Cambridge.

Minor products

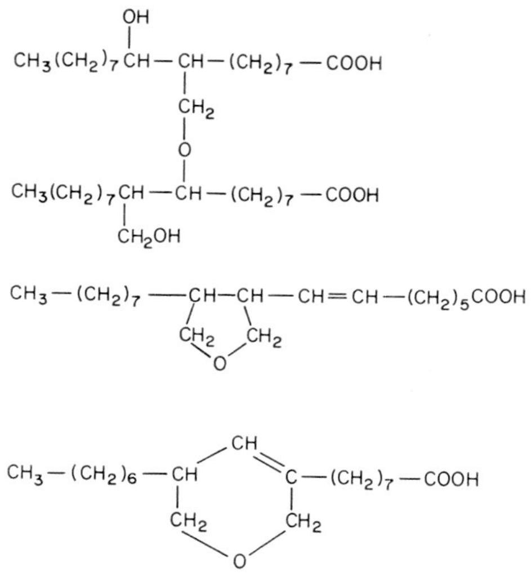

$$CH_3(CH_2)_7CH(OH)-CH(CH_2-O-)-(CH_2)_7-COOH$$

$$CH_3(CH_2)_7CH(CH_2OH)-CH(-O-)-(CH_2)_7-COOH$$

$$CH_3-(CH_2)_7-\underset{CH_2-O-CH_2}{CH-CH}-CH=CH-(CH_2)_5COOH$$

$$CH_3-(CH_2)_6-CH\,(-CH=C\,(-CH_2-O-CH_2-))-(CH_2)_7-COOH$$

Introduction

The ideal fixative should preserve a tissue against microbial activity, osmotic damage and autolysis, and should, of course, ensure that even after removal of the fixative the structure of the tissue remains an accurate representation of the structure in the living tissue; additionally, cells in the tissue should retain their original size, no material should escape from the cells during the fixation and the reactivity of the substances in the cell, as chemicals, should remain sufficiently high to enable them to be located and, wherever possible, identified by means of suitable reactions. Preferably, the fixative should indissolubly link together protein, lipid and other cellular constituents in the form of a macromolecular network, but without affecting subsequent reactions of a histochemical nature by destroying functional groups.

This, then, is the ideal fixative. However, in this context as in many others, the search for ideality can be equated in terms of reality with the search for the Philosophers Stone; it seems likely that we must search for the most ideal of the non-ideal fixatives amongst the aldehyde fixatives, of which the three most interesting examples, as viewed in the light of present-day knowledge, are formaldehyde, acrolein and glutaraldehyde.

Formaldehyde is known to fulfil many of the criteria of the ideal to at least a satisfactory extent; thus, it is a 'polymeric' fixative and converts cytoplasmic protein into an insoluble macromolecular network, through its ready reaction with proteins. This has hitherto been considered as of extreme importance in fixation and unfortunately little attention has been paid to the reaction of formaldehyde with other cellular components, though fixation of other classes of compounds present in tissues may be as important as fixation of proteins.

Formaldehyde, being the simplest member of a homologous series, the saturated aliphatic aldehydes, has of course many reactions peculiar to itself, but is also the most reactive member of the series; on grounds of reactivity, acrolein ($CH_2=CH-CHO$), the simplest member of the series of $\alpha\beta$-unsaturated aldehydes, is the only aldehyde which might reasonably be expected to mimic the actions of formaldehyde as a fixative (Jones, 1969*a*,*b*). Acrolein is, in some respects, more reactive than formaldehyde, and has been suggested for the tanning of hides (Gustavson, 1940; Nozaki *et al.*, 1953); with proteins it behaves like formaldehyde, linkages occurring at -NH- groups, to give initially

$$\begin{array}{c} \qquad\qquad\qquad\quad | \\ CH_2{=}CH{-}CH{-}N{-} \\ \qquad\;\; | \\ \qquad\;\; OH \end{array}$$

followed rapidly by a polymerization process (van Winkle, 1962). It is also known to react with fatty acids (Hall & Stern, 1955), giving esters of β-hydroxypropionaldehyde:

$$R{-}COOH \xrightarrow{CH_2{=}CH{-}CHO} R{-}COO{-}CH_2CH_2CHO$$

The resulting aldehyde product is theoretically capable of reacting with proteins and a mechanism thus exists for linking non-esterified fatty acids to proteins.

Acrolein is capable of undergoing a Diels-Alder reaction (Diels & Alder, 1928), reacting either as a diene or a dienophile (Whetstone *et al.*, 1951); the reaction as a diene could have an application in the 'fixation' of unsaturated fatty acids:

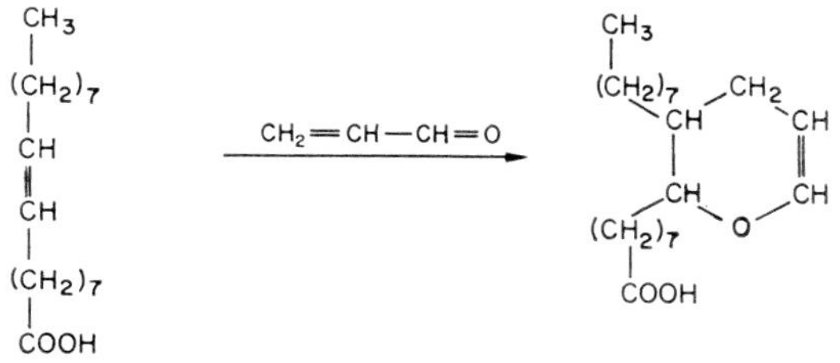

$$\begin{array}{c} CH_3 \\ | \\ (CH_2)_7 \\ | \\ CH \\ \| \\ CH \\ | \\ (CH_2)_7 \\ | \\ COOH \end{array} \xrightarrow{CH_2{=}CH{-}CH{=}O} \begin{array}{l} CH_3 \\ | \\ (CH_2)_7 \quad CH_2 \\ \quad\; CH \qquad\quad CH \\ \quad\;\, | \qquad\qquad \| \\ \quad\; CH \qquad\quad CH \\ (CH_2)_7 \quad O \\ | \\ COOH \end{array}$$

This reaction is not as significant as it may appear, since it requires forcing conditions and the product of the reaction is not very reactive; however, hydrolysis of the reaction product gives rise to another containing a side-chain aldehyde group

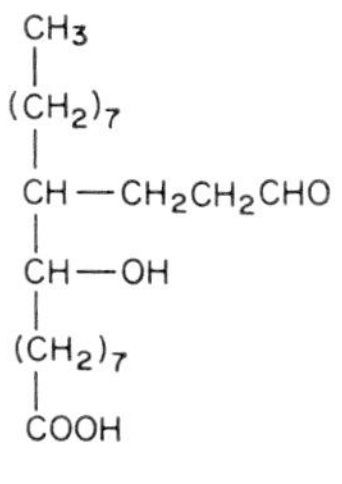

$$\begin{array}{l} CH_3 \\ | \\ (CH_2)_7 \\ | \\ CH{-}CH_2CH_2CHO \\ | \\ CH{-}OH \\ | \\ (CH_2)_7 \\ | \\ COOH \end{array}$$

which is not only reactive, but detectable by histochemical methods, e.g. with Schiff's reagent or 2,4-dinitrophenylhydrazine. It is unfortunate that conjugated dienoic acids do not normally occur in biological material, as they also react with acrolein (behaving as a dienophile) to give derivatives with reactive centres. On the above basis acrolein would thus appear to have a number of properties contributory to its use as a 'lipid' and general fixative.

Glutaraldehyde is one of the newer fixatives to appear on the scene; Gustavson (1940) did not include it in his study of tanning powers of aldehydes, although he did comment that tanning power was possessed only by aliphatic aldehydes. Recent work has demonstrated that glutaraldehyde is an excellent general fixative; Hündgen (1968) ranked it and acrolein as the best fixatives, better than formaldehyde for morphological fixation and enzyme preservation. Hopwood (1967), however, found that glutaraldehyde inactivates enzymes more rapidly than formaldehyde, and commented on the presence of demonstrable carbonyl groups in tissues after fixation. In a subsequent study in which he compared glutaraldehyde with formaldehyde and α-hydroxyadipaldehyde, Hopwood (1969*a*,*b*) confirmed that glutaraldehyde cross-links protein during fixation with much more ease and rapidity than other aldehydes. Gigg & Payne (1969) and Roozemond (1969*b*) have also shown that glutaraldehyde cross-links phosphatidyl ethanolamine to protein during fixation.

The motivation behind the study described in this paper was mainly that of placing formaldehyde fixation on a more reliable scientific footing with respect to lipids, and to build up a model of aldehyde reactions with unsaturated fatty acids that could be applied to other aldehydes. Most attention has been paid to formaldehyde, since it is in general use, and technical difficulties associated with the study of the reaction were easily surmountable. The study performed with acrolein emphasized the technical difficulties. A study with glutaraldehyde of the type described here would be impossible.

The reactions of formaldehyde with proteins are numerous and well authenticated; formaldehyde can, and does, combine with many of the groups found in proteins. The major functional groups which formaldehyde reacts with, and the amino acids or situations in which they are found, are shown below (from Pearse, 1960).

Functional group	*Situation* (examples only)
$—NH_2$	Arginine, lysine or terminal amino acids in peptide chains
=NH	Arginine, histidine, hydroxyproline, proline, tryptophan
—C=O | —N—H	Peptide bonds
—OH	Hydroxy-proline, serine, threonine
—SH	Cysteine
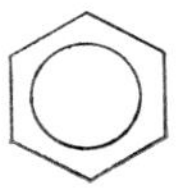	Phenylalanine, tyrosine, tryptophan

In aqueous solution, formaldehyde behaves and reacts in a hydrated form. In acid conditions, however there are higher concentrations of the more reactive electrophile

(III) which may be considered to be derived from the carbonyl form (I) or from the hydrated form (II) by protonation.

$$\underset{(I)}{H_2C{=}O} \;\xrightleftharpoons{H_2O,\ fast}\; \underset{(II)}{H_2C(OH)_2}$$

(II) $\xrightarrow{H^+(H_3O^+)\ [protonation]}$ $\left[H_2\overset{+}{C}(OH_2)(OH) \longleftrightarrow H_2C(\overset{+}{O}H_2)(OH) \right]$ $\xrightarrow{-H_2O}$ (III)

(I) $\xrightarrow{H^+(H_3O^+)\ [protonation]}$ (III)

$$\underset{(III)}{H_2\overset{+}{C}{-}OH \longleftrightarrow H_2C{=}\overset{+}{O}H}$$

I→III is the more likely of these two mechanisms (De La Mare & Bolton, 1966); in the carbonyl form, reaction is possible because of the inductive effect. The extent of the above reaction(s) increases as the pH of the solution falls. The carbonium ion ($^+CH_2OH$) produced is a reactive electrophile, and is responsible for the majority of the 'fixative' effects of formaldehyde, so that those workers who have suggested the use of buffered formaldehyde-containing fixatives are, therefore, merely suggesting methods of reducing the fixative powers of formaldehyde.

The species $^+CH_2OH$ reacts at electron-rich centres, e.g. at double bonds, or with nitrogen, oxygen or sulphur atoms (the electronegativity or electron-attracting powers of these atoms causes them to be relatively 'rich' in electrons). The reaction with alkenic double bonds is of particular interest in fixation; fats of biological origin always contain such double bonds. For example, the fat in the adipose tissue of ruminant animals, normally considered the most saturated fat of animal origin, has an average iodine value of about 40. This means an average of about 1.5 double bonds per molecule of triglyceride; there can be very few molecules of triglyceride present in this adipose tissue which are completely saturated. Any reaction of formaldehyde with double bonds will, therefore, require evaluation to determine both the role it may play in fixation and storage of lipid-containing tissues, and what, if any, modification of the histochemical properties may occur. Prins (1919) described a reaction which is in keeping with the postulated reactive species, namely $^+CH_2OH$. The reaction has been confirmed and extended by, *inter alia*, Olsen & Padberg (1946), Olsen (1946, 1947, 1948), Smissman & Mode (1957), Smissman & Witiak (1960), Blomquist & Wolinsky (1957), Blomquist *et al.* (1957*a*,*b*), Dolby (1962), Dolby *et al.* (1963), and Le Bel *et al.* (1963).

The Prins reaction with an alkene takes place readily under acidic catalysis:

(*a*) —Initially the carbonium ion ‘embeds’ itself in the π-orbitals of the double bond to form a π-complex.

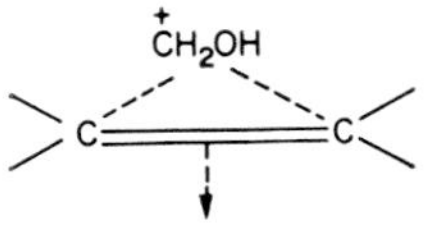

(*b*) —A σ-bond is formed with one or other of the carbons. Overall structure of the alkene governs which carbon forms this bond.

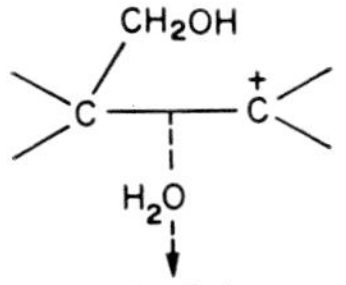

(*c*) —A proton is released in this phase of the reaction to regenerate the proton consumed in the formation of the $^{+}CH_2OH$ carbonium ion.

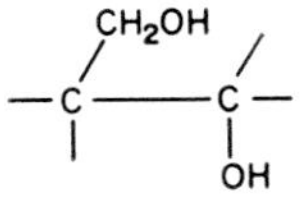

The compound produced in the reaction is a 1,3-glycol. However, if in step (*c*) above, formaldehyde reacts instead of water, a 1,3-dioxan may be produced instead:

(*d*) —Electrophilic attack of the carbonium ion on the carbonyl oxygen.

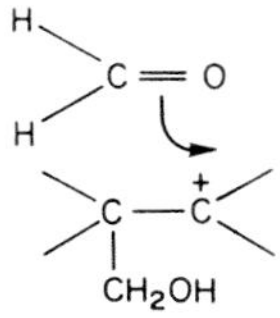

(*e*) —Intramolecular electrophilic attack of a carbonium ion on the electronegative oxygen atom in the hydroxymethyl group. Once again a proton is released.

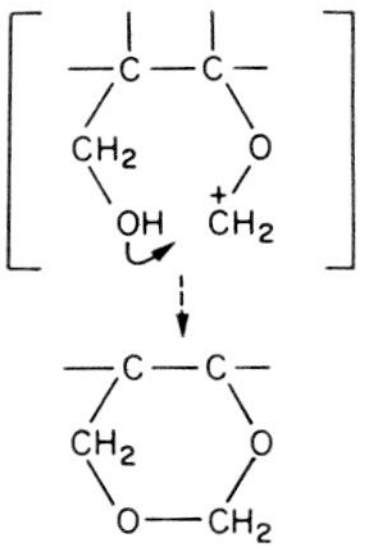

Steps (*d*) and (*e*) in the reaction are of interest for ‘fixation’ for obviously here is a mechanism whereby adjacent fatty acid molecules may be joined by chemical bonds. Thus, if an *inter*-molecular electrophilic attack occurs in step (*e*) above, instead of the intramolecular attack, fatty acid dimers may be produced, and given sufficient time, the reactions may progress to yield a low mol. wt. polymer.

The reactions shown above have been verified for alkenes, but there is little doubt that the immediate environment of the double bond will affect the reaction only in detail, not in principle, and it is difficult to understand why the possibility of a reaction between formaldehyde and unsaturated fatty acids has only been discussed on two occasions previously in the literature. Wolman & Greco (1952) stirred oleic acid and formaldehyde together for considerable lengths of time and proceeded, using standard analytical techniques, to demonstrate the presence of carbonyl groups. As they took no precautions to exclude oxygen during the reaction, there is a high probability that some oxidation had occurred at the double bond and the allylic methylene groups, thus giving rise to free aldehydes. Their suggestion that side-chain carbonyl groups were formed is obviously not in keeping with the accepted mode of reaction of formaldehyde. More recently, Artun (1966) carried out a classical Prins reaction with oleic acid, using 95% acetic acid as solvent, sulphuric acid as proton donor, and paraformaldehyde to supply formaldehyde, but he did not attempt to separate the individual components of the reaction mixture.

The following aspects of the Prins reaction may bear on the reaction of formaldehyde with unsaturated fats in tissues:

1. The reaction is, at least in the early stages, reversible, and will attain an equilibrium, whose rate of attainment will depend on the temperature and concentration of the reactants (as this is probably a second order reaction, the time is inversely proportional to the initial concentration of reactants). The temperature during fixation is normally low, so the reaction will be slow on this account but the concentration of formaldehyde during normal fixation is (or should be) high compared to that of the material fixed; a piece of tissue in formaldehyde solution represents, however, a complicated physicochemical system which it would be wrong to analyse accurately.
2. The formation of the carbonium ion, $^{+}CH_2OH$, will not be a rate-determining step, as this is a fast reaction, *provided* the fixative is sufficiently acid for there to be a reasonable concentration of protons to take part in the reaction.

The Prins reaction, or a variation of it, appears to be the only reaction of unsaturated fatty acids that could take place during fixation. Following the motivation outlined above, it was, therefore, decided to investigate this particular reaction in detail; the study described in this paper may be considered as divisible into the following parts:

(1) The demonstration that formaldehyde and acrolein react with pure unsaturated fatty acids under the 'forcing' conditions of elevated temperature and high concentration to give a number of products.

(2) The demonstration that formaldehyde and acrolein react with pure unsaturated fatty acids and triglycerides under conditions identical with those of histological fixation to yield products identical to those in (1).

(3) The isolation and characterization of the new products formed in the reactions, and a brief kinetic study of their formation.

(4) The demonstration that the reactions occurring during fixation alter the 'fatty acid pattern' of biological materials significantly.

Unesterified fatty acids were generally used in preference to esterified fatty acids, as they reacted more rapidly, were easier to process and were generally of greater initial purity. However, some experiments were performed with pure triglycerides to confirm that the reaction sequence for the aldehydes and non-esterified fatty acids also applied to esterified fatty acids.

Materials, methods and results

Materials

The studies described were performed with 'A' grade (Calbiochem, Koch-Light or

Sigma) palmitic, stearic, oleic, elaidic and linoleic acids, and glyceryl trioleate and biochemical grade linolenic acid (BDH Ltd). Normal laboratory grade 1- and 2-octene (BDH Ltd) were used for testing only.

The fixatives used in the experiments, with their composition, are shown in Table 1.

Table 1. Fixatives used in experiments

Fixative	*Abbreviation*	*Composition/100 ml*		*pH values**				
		Aldehyde[1]	*Ionic*[2]	*A*	*B*	*C*	*D*	*E*
Formalin-saline	F or FS	10 ml 40% formaldehyde solution	0.9 g Nacl	4.1	4.8	4.8	4.85	4.15
Acrolein-saline	Aa or AS	4 ml acrolein	0.9 g Nacl	4.4				
Acrolein-saline, buffered	Ab	4 ml acrolein	0.9 g Nacl 0.264 g disodium hydrogen phosphate 0.136 g sodium dihydrogen phosphate ($2H_2O$)	7.0				
Buffered formalin	BFS	10 ml 40% formaldehyde solution	0.44 g disodium hydrogen phosphate 0.22 g sodium dihydrogen phosphate ($2H_2O$)	7.2	6.9	6.85	6.75	5.70
Formalin-calcium acetate	FCA	10 ml 40% formaldehyde solution	1.6 g calcium acetate	7.1	6.35	6.0	6.1	—

Notes: *pH values. 15–20 gm of mixed tissues (from ferrets) were fixed in 250 ml of each fixative. A = before fixation; B = after 24 hr fixation; C = after 7 days fixation; D = after 28 days fixation; E = after 1 year or more fixation.

[1] Acrolein stabilized with hydroquinone.

[2] Weights given as anhydrous material unless otherwise specified.

Methods

The conditions for reactions are noted in the appropriate sections. After reaction, substrates were normally extracted with petroleum spirit (b.p. range 40–60°C); the solutions in petroleum spirit were always washed with water and dried over anhydrous sodium sulphate. They were then evaporated to dryness in a stream of nitrogen, and methylated, if present as free fatty acids, by adding an ethereal solution of diazomethane, prepared according to Vogel (1957), to a solution of the acid in ether until a yellow colour persisted. Excess reagent and solvent were then removed by evaporation in a stream of nitrogen, leaving the mixture of methyl esters in a small bulk for storage or analysis. Triglycerides were methylated by refluxing with 5 ml 5% sulphuric acid in methanol for 1 hr in tubes, sealed at one end, approximately 25 cm long and 1.5 cm bore (Bowyer *et al.*, 1963). The methyl esters were extracted with petroleum spirit after dilution of the methylating mixture with water. The extract was washed with water and dried over a sodium sulphate-sodium bicarbonate mixture (4:1 w/w) and the solvent removed by evaporation as above or under a mild vacuum.

Aliquots of the methyl ester solutions were analysed on either a Pye Gas Chromatograph (having a conversion to count ^{14}C as well as a mass detector) or a Pye Model 124 Gas Chromatograph. Column materials used were 10% polyethylene-glycol adipate (PEGA) on Celite, 10% Apiezon L (APL) on Celite, and 15% free fatty acid phase (FFAP) on Celite. The Celite was in all cases 100–120 mesh. The column length used in the Pye Argon machine was 3 ft, and in the Model 124 normally 5 ft, though a 7 ft preparative scale column was also used in this machine. Columns were always used under isothermal conditions. Compounds were characterized on each type of column by means of their carbon numbers; standard mixtures containing the methyl esters of myristic ($C_{14:0}$), palmitic ($C_{16:0}$), stearic ($C_{18:0}$) and behenic acid ($C_{20:0}$) were run on each column immediately before a batch of experimental samples. The methyl esters of the standard fatty acids were allotted carbon numbers of 14:0, 16:0, 18:0 and 20:0 respectively. The logarithms of their retention times were plotted against the carbon numbers assign-

Table 2. % composition by mass of 'oleic acid' before and after treatment with formaldehyde at 100°C.

Treatment[1]		*Approximate Carbon Nos. on PEGA*							
		16.0	18.0	18.3	18.9	23.35	23.85	24.15	25.9
Reagent blank		—	—	—	—	—	—	—	—
Palmitic acid, no treatment		98.67	—	0.24	—	—	—	—	—
Palmitic acid (H)[2]	8 hrs	99.14	—	—	—	—	—	—	—
Oleic acid, no treatment		0.19	—	98.84	—	—	—	—	—
Oleic acid (C)[3]	8 hr	0.22	—	98.91	—	—	—	—	—
Oleic acid (H)	2 hr	T[4]	—	86.64	—	3.43	1.05	4.44	1.90
Oleic acid (H)	6 hr	0.39	—	75.97	0.35	2.92	2.29	5.77	10.71
Oleic acid (H)	6 hr	0.22	—	79.01	—	2.52	0.96	3.95	10.94
Oleic acid (H)	8 hr	0.33	—	64.21	—	5.49	—	9.14	14.52
Oleic acid (H)	10 hr	1.12	—	57.92	—	14.2	—	20.92	—

Notes: [1] Some of the 'treated' samples contained short chain fatty acids as breakdown products. These were measured, but not included in this table.
All samples were chromatographed as their methyl esters on a Pye Argon Chromatogram using a PEGA column.
[2] H = 40% formaldehyde, 10% acetic acid, 5% hydrochloric acid, pH approximately 0, at 100°C
[3] C = 10% acetic acid, 5% hydrochloric acid, pH = 0.5, at 100°C.
[4] T = trace.

ed; this gave a straight line from which carbon numbers of unknowns could be calculated. Other experimental details are given in the relevant sections. The alkenes used for comparative studies were not methylated, but were chromatographed as the extracted mixtures.

Demonstration of the reactions of formaldehyde and acrolein with unsaturated fatty acids

1. *Reaction under forcing conditions:*
15–50 mg samples of oleic acid were heated for up to 20 hr with 10 ml of a reagent solution composed of 6 M aldehyde (acrolein or formaldehyde) dissolved in acetic acid (2 M)-hydrochloric acid (0.5 M)-water. The pH of this reagent was approximately 0. For control purposes, reagents, without added oleic acid, were treated identically, and 50 mg samples of oleic acid were also heated for up to 20 hr with reagent solutions consisting only of acetic acid-hydrochloric acid-water. For comparison the reaction of formaldehyde with other fatty acids (see *Materials*) was also studied. After reaction, fatty acids were extracted, methylated, and the mixtures of methyl esters were analysed by gas chromatography. Some mixtures were also examined by i.r., proton resonance and mass spectrometry.

The results obtained with oleic acid and formaldehyde were satisfactory as regards observing new compounds; up to seven new products could be demonstrated (Table 2), the mixture being correspondingly more complex with linoleic and linolenic acids (Figs. 1 & 2, Tables 3 & 4). Blanks were satisfactory; reagent blanks showed no peaks on

Table 3. % composition by mass of 'linoleic acid' before and after treatment with formaldehyde at 100°C.

Approximate Carbon Nos. on PEGA	*Description of treatment:* None	*Heating with mixture H, 8 hr*	*Heating with mixture H, 10 hr*
16.0 and under	T	T	T
18.9	99.09	74.96	64.87
21.0 to 23.0	—	—	0.95 (3)[1]
23.4	—	6.04[2]	6.89
24.3	—	1.71	1.48
24.6	—	2.31	2.47
24.9	—	—	0.41
25.3	—	—	1.23
26.1	—	11.44	12.03
Over 26.1	—	—	8.99 (3)[1]

Notes: [1] Figures in brackets represent no. of compounds observed in that range of carbon nos.
[2] This compound had a double peak, neither constituent of which could be measured separately.
Chromatography and abbreviations as indicated in the Notes of Table 2.

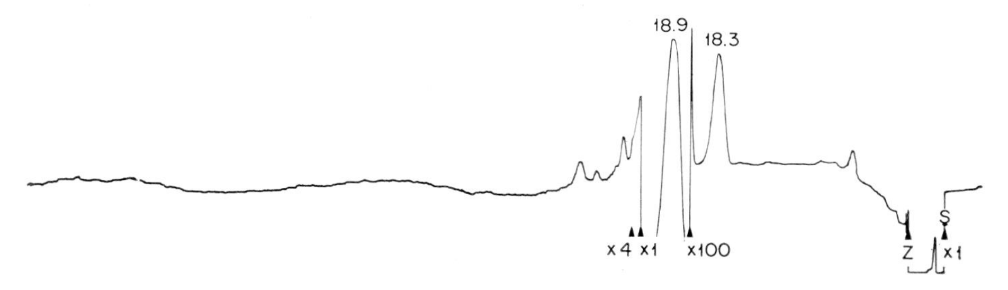

Figure 1. Linoleic acid (18.9), chromatographed as the methyl ester in a Pye argon chromatograph PEGA column, 190°C.

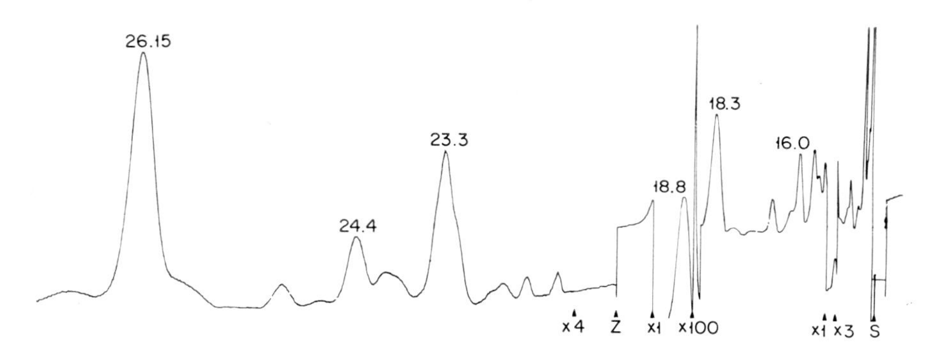

Figure 2. Linoleic acid, after 6 hr treatment with 40% formaldehyde —HCl—acetic acid. Cf. with Fig. 1. Chromatographed as methyl esters in a Pye argon chromatograph, PEGA column, 190°C. Z represents a base line change.

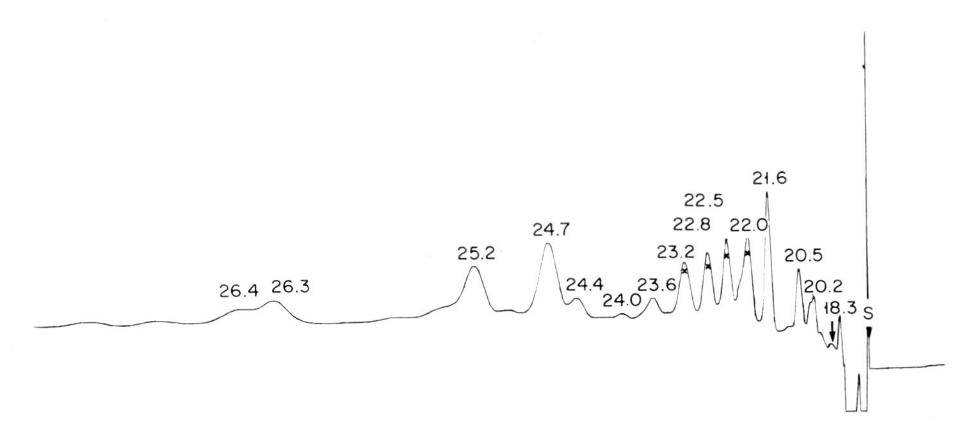

Figure 3. Oleic acid after reaction with acrolein under forcing conditions for 12 hr chromatographed as methyl esters in a Pye model 124 chromatograph, FFAP column. The peaks marked X are acrolein polymers, present also in the blank.

Table 4. % composition by mass of 'linolenic acid' before and after treatment with formaldehyde at 100°C.

Approximate Carbon Nos. on PEGA	*Description of treatment:* *None*	*Heating with mixture H, 8 hr*
Under 16.0	0.32 (5)[1]	0.48 (3)[1]
16.0	1.23	31.67
16.4 to 17.4	0.16 (3)[1]	0.86 (1)[1]
18.0	0.81	22.56
18.3	3.99	—
18.9	18.57	3.66
19.6	73.97	13.57
20.0 to 22.9	0.77 (7)[1]	2.78 (7)[1]
23.3	0.19	4.57
23.9	0.19	1.24
24.3	—	4.25
24.6 to 25.3	—	1.36 (3)[1]
26.1	—	8.10
26.3	—	4.85
Over 26.3	—	0.06

Note: [1] Figures in brackets represent no. of compounds observed in that range of Carbon Nos.
H, see footnote 2, Table 2.

Table 5. % composition by activity of [^{14}C]oleic acid before and after treatment with formaldehyde at 100°C.

Description of treatment:	*Carbon Nos. on PEGA columns:* 16.0	18.0	18.3	20.0	20.6	23.35	24.15
(1) Oleic acid	T		99.0	T	T		
(2) Oleic acid (H) 8 hr	T	—	76.0	T	T	7.4	16.5
(3) Oleic acid (H) 8 hr	T	—	82.8	T	T	7.9	9.3
(4) Oleic acid (H) 8 hr	T	—	87.5	T	T	2.7	9.8
(5) Oleic acid (H) 8 hr	T	—	81.1	0.6	0.4	8.1	9.7
Mean ± S.D. of (2)–(5) inclusive	—	—	81.8±4.7	—	—	6.5±2.6	11.3±3.5

Abbreviations: H, T — see footnotes 2 & 4 of Table 2 respectively.

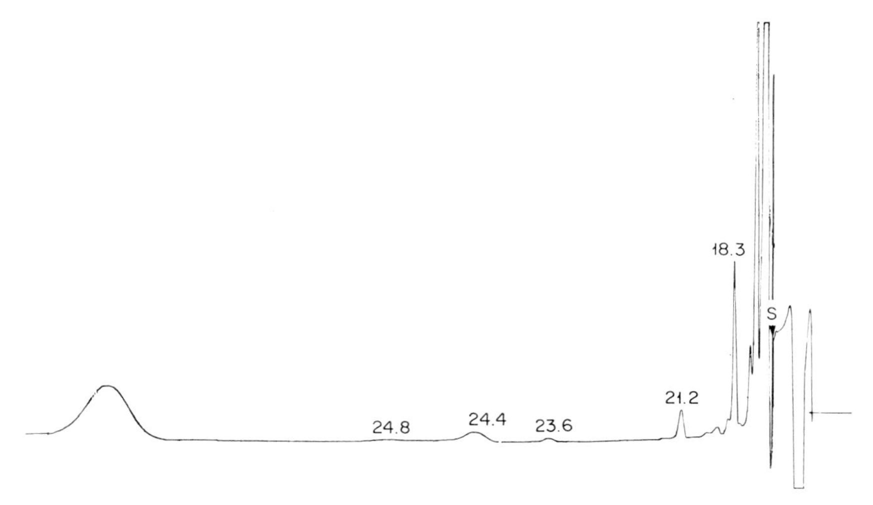

Figure 4. Oleic acid after reaction with formaldehyde under forcing conditions for 12 hr, chromatographed as methyl esters in a Pye model 124 chromatograph, FFAP column.

gas chromatography and the palmitic acid control showed no peaks other than those expected. Results obtained with [1-^{14}C]oleic acid were very similar to those obtained with the unlabelled acid (Table 5).

I.r., proton resonance and mass spectra provided evidence of reaction, but the results obtained by use of these techniques on the reaction mixtures were not as useful as expected. In common with the gas chromatography results, they mainly confirmed only that reactions had occurred, though they also gave some clues as to the nature of the functional groups occurring in molecules after reaction.

Considerable carbonization occurred in the acrolein reaction mixtures. The experiments were not performed quantitatively, but very little oleic acid remained after 12 hr reaction with acrolein under these conditions. Representative results are shown in Fig. 3 and comparison results with formaldehyde in Fig. 4. Analyses by other procedures were not performed on acrolein reaction mixtures.

2. *The 'fixation' reaction*

(a) *With pure acids.* 5 mg portions of stearic (0) oleic (1) elaidic (1*a*) and linoleic (2) acids were fixed in 10 ml portions of unbuffered formalin-saline (*F*), unbuffered acrolein-saline (*Aa*), or buffered acrolein-saline (*Ab*). At least five samples of each combination, in 15 ml sample vials, were shaken continuously in a mechanical shaker for 36 days, after which fatty acids were extracted, methylated, and analysed by gas chromatography (FFAP columns only). Controls consisted of fixatives alone or of fatty acids shaken with 0.9% saline under identical conditions. In all cases, controls were treated exactly as the experimental samples, including extraction, methylation and analysis.

Calculation of results was restricted to a determination of the change in purity of the original material fixed, as this was considered a valid parameter of reactions occurring. Some representative gas chromatograms are shown in Figs. 5–10 and Table 6 gives an

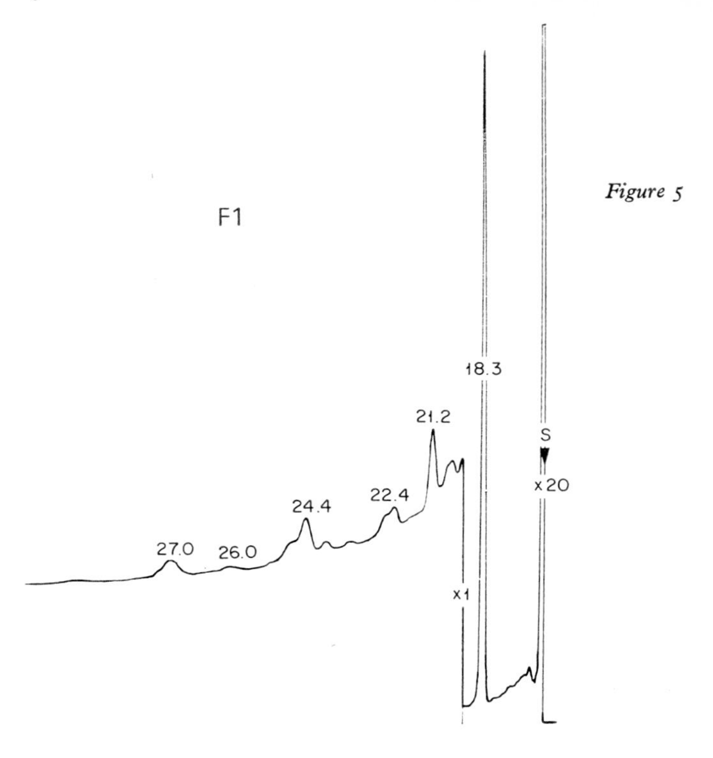

Figure 5

Figures 5 & 6. Oleic acid (18.3). These figures show typical signs of the 'fixation reaction'.

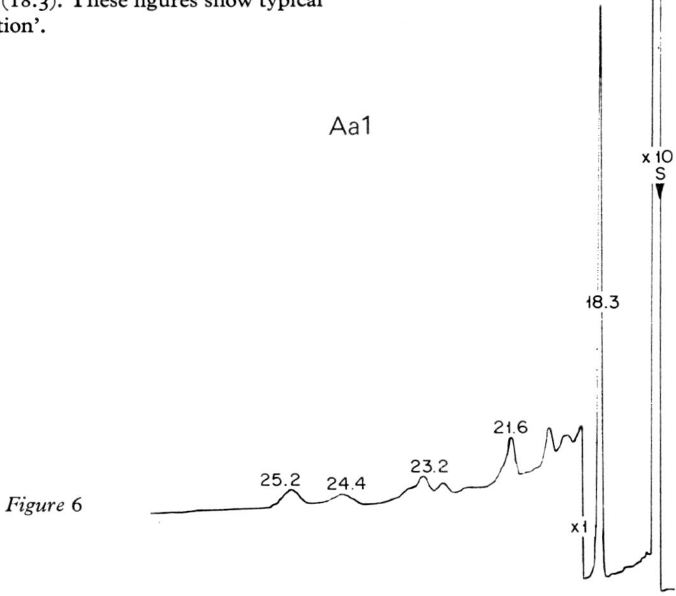

Figure 6

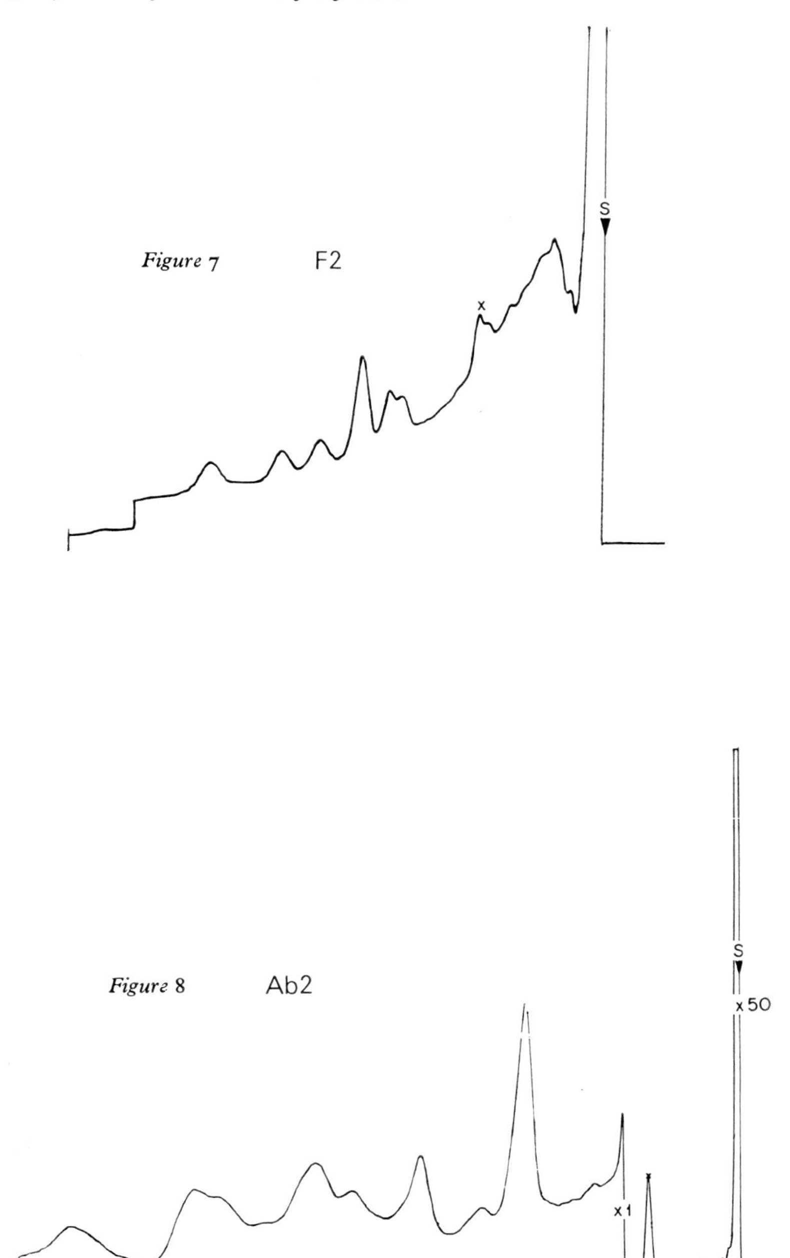

Figures 7, 8 & 9. Linoleic acid (18.9; X). There are considerable changes attributable to the 'fixation' reaction present. Linoleic acid 'fixed' in formalin saline shows the greatest change (Fig. 7), while the change attributable to 'fixation' in acrolein-saline appears relatively minor. The reason for this is not apparent.

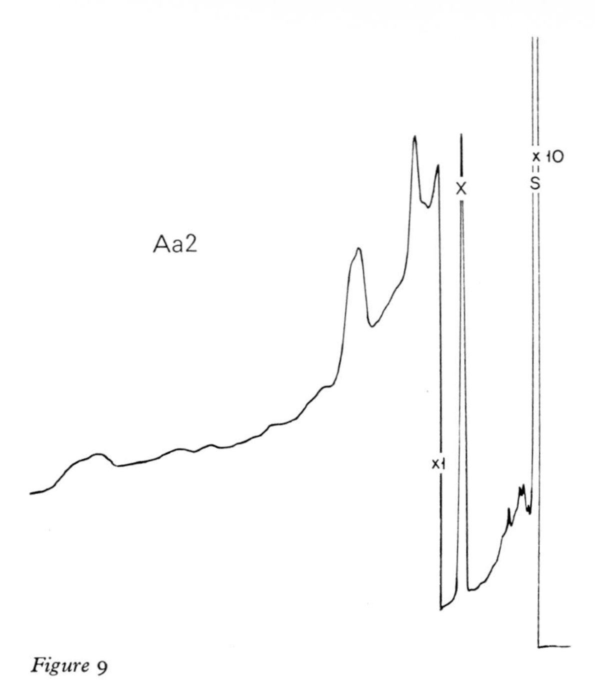

Figure 9

overall summary of the changes seen. The results indicate that oleic acid, glyceryl trioleate and linoleic acid all react significantly with acrolein and formaldehyde, but elaidic acid does not. An effect with stearic acid was neither expected nor seen, but changes in 'trace' fatty acids originally present as impurities were observed.

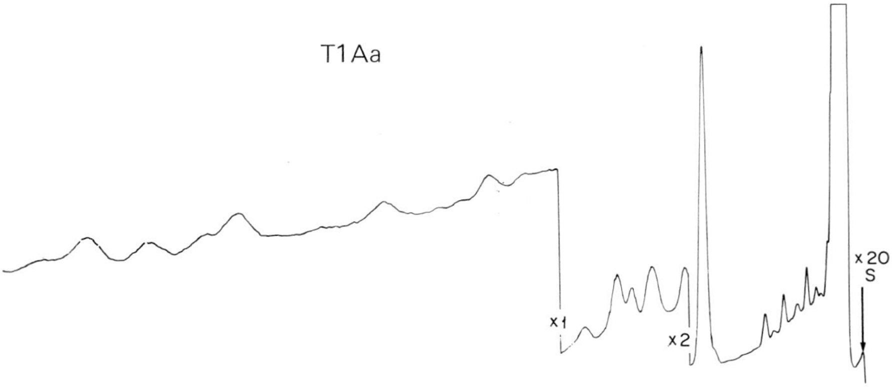

Figure 10. Triolein 'fixed' in acrolein-saline. The 'fixation change' produced here is relatively minor, consisting of production of a number of compounds with higher carbon number. The large peak immediately before the $\times 2$ sensitivity change is methyl oleate.

Table 6. The effect of formaldehyde and acrolein on unsaturated fatty acids and neutral lipids. Results expressed as % of fatty acid remaining after 36 days' fixation, mean ± S.D. (five observations).

	Before fixation (1)	*After fixation in:* *formalin-saline* (2)	*acrolein-saline* (3)	*buffered acrolein-saline* (4)	*Significance of results* (*t*-test)
Stearic acid	97.2±1.1	97.8±0.84	96.04±0.59	96.3±1.96	No significant difference.
Oleic acid	97.8±1.08	92.1±2.07	91.6±0.84	91.9±1.24	2, 3 and 4 significantly reduced from 1 (0.1% level).
Elaidic acid	59.6±3.34	54.9±4.06	55.5±2.91	56.9±2.84	No significant difference.
Linoleic acid	96.5±2.48	43.4±5.06	50.4±6.35	36.8±9.33	2, 3 and 4 significantly reduced from 1 (0.1% level).
Triolein	98.3±0.57	95.0±0.53	95.0±0.69	95.4±0.69	2, 3 and 4 significantly reduced from 1 (0.1% level).

(b) *With 'artificial' tissues.* A solid medium was made up to contain 10% gelatine, 2% agar, 1% sucrose, dissolved in boiling water and allowed to cool in Petri dishes. Pieces of approximately 0.7 cm diameter and 0.5 cm thickness were cut out with a cork borer and freeze-dried. 10 μl of liquid fatty acid was placed on each piece and was rapidly absorbed. These 'tissues' were then fixed for 14 days in different fixatives. At the end of this time, both 'tissue' and fixative were extracted (by homogenizing with chloroform) and prepared for gas chromatography as described previously. As controls, pieces of freeze-dried 'tissue' without fatty acid were fixed and extracted likewise.

The changes observed were comparable to those seen under (2*a*) above; representative results, for linoleic acid, are given in Table 7.

(c) *With 'genuine' tissues.* The types of fatty acids present in a sebaceous adenoma and a sebaceous carcinoma were determined before and after fixation for two years; fatty acid analyses of the fresh materials were originally performed to confirm their tissue of origin (Essenhigh *et al.*, 1964). Tissue not analysed at that time was stored in buffered formalin, the pH of which fell to about 5.0 after three months. After two years fixation, further tissue was removed and analysed for fatty acids; tissues were homogenized with chloroform: methanol (2:1, v/v), the extracts were purified by washing with 0.9% saline (Folch *et al.*, 1957) and lipids were recovered by evaporation of the solvent. The lipids were methylated *in toto* with methanol containing 5% sulphuric acid, and methyl esters were purified and chromatographed as described above.

Results are presented in Tables 8 & 9. The fatty acid patterns of both tissues are unrecognizable after two years' storage in formaldehyde.

Table 7. % composition by mass of 'linoleic acid' before and after fixation in formaldehyde-based fixatives. Chromatographed as methyl esters on a Pye Argon chromatograph—PEGA column.

Description of treatment:	*Approximate Carbon Nos. on PEGA columns* *Under* 18.0	18.3	18.9	20.22.5	24.0	24.3	26.15
(1) Linoleic acid	0.55	0.36	99.09	—	—	—	—
(2) Linoleic acid shaken with water for 4 days	0.1	0.92	98.98	—	—	—	—
(3) Linoleic acid fixed in FS, pH 4.0, for 4 days	0.1	1.51	93.32	—	0.89	1.20	—
(4) Linoleic acid fixed in FCA, pH 7.0, for 4 days	0.19	1.08	95.15	—	—	1.47	—
(5) Linoleic acid fixed in mixture B, pH 9.0, for 4 days	0.39	1.23	87.02	1.81	1.19	3.37	—
(6) Linoleic acid* fixed for 14 days in FS, pH 4.0	18.64	T	74.56	2.48	4.13	0.19	—

Abbreviations: FS, FCA—see Table 1. T, see footnote 4, Table 4.
B = 3% sodium borate ($Na_2B_4O_7$) in 40% formaldehyde solution.
* signifies [^{14}C]linoleic acid, estimated by counting rate, not by mass.

Table 8. % composition by mass of methyl esters of the fatty acids in a sebaceous adenoma before and after fixation in 10% buffered formalin (pH 5.0) for 2 years.

Description	*Before or after fixation*	*Approximate Carbon Nos. of the Methyl Esters on PEGA* *Under* 16.0	16.0 *to* 19.4 *saturated*	*unsaturated*	*Over* 19.4	*Ratios* 16.0 (16:0) / 16.3 (16:1)	18.0 (18:0) / 18.3 (18:1)
Cholesterol esters	Before	16.3	19.6	57.2	6.9	0.81	0.26
	After	27.0	16.6	43.1	13.3	1.19	0.39
Triglycerides	Before	19.0	41.2	35.8	3.8	4.62	0.31
	After	23.1	44.0	24.5	6.4	7.23	0.36
Free fatty acids	Before	20.8	31.8	43.4	4.1	2.67	0.26
	After 1 yr†	44.6	23.8	31.6	—	1.93	0.27
	After 2 yr†	62.4	11.8	19.6	6.1	1.76	0.25

Notes: †There were two free fatty acid fractions after fixation. Fraction 1 was least polar on thin-layer chromatography.
Acids with Carbon Nos. less than 12 are not detectable by this method.

Table 9. % Composition by mass of methyl esters of the fatty acids in a sebaceous carcinoma before and after fixation in 10% buffered formalin (pH 5.0) for 2 years.

Description	*Approximate Carbon Nos. of the Methyl Esters on PEGA*† *Before or after fixation*	*Under* 16.0	16.0 *to* 19.4	*Over* 19.4	*Ratios* 16.0 ($C_{16:0}$) / 16.4 ($C_{16:1}$)	18.0 ($C_{18:0}$) / 18.3 ($C_{18:1}$)
Cholesterol esters	Before	42.96	32.03	25.03	1.76	0.36
	After	6.47	65.88	27.65	2.63	0.18
Triglycerides	Before	7.65	62.99	29.36	2.52	0.38
	After	1.71	86.42	11.87	7.01	0.52
Free fatty acid	Before	39.61	40.98	19.41	2.62	0.46
	After	2.53	89.85	7.60	3.39	0.61
Lecithin	Before	12.68	66.27	21.04	4.01	0.42
	After	5.73	84.96	9.31	4.01	0.50
Lysolecithin	Before	26.78	29.39	43.83	1.13	0.37
	After	5.44	75.86	18.72	3.33	0.64
Sphingomyelin	Before	16.68	25.55	57.80	2.54	4.98
	After	2.48	58.71	38.83	26.78	0.77

Notes: †Acids with Carbon Nos. less than 12 are not detectable by this method.

(d) *Separation and identification of the products from the reaction of oleic acid with formaldehyde.* 2 g oleic acid was refluxed for 24 hr with a reaction mixture composed of formaldehyde (50 ml 40% solution), glacial acetic acid (45 ml), and hydrochloric acid (5 ml, concentrated, analytical reagent grade). The reaction mixture was then diluted to 1 litre with distilled water, and products extracted three times successively with petroleum spirit (b.p. range 40–60°C). The petroleum extracts were washed with water, dried over sodium sulphate and evaporated to dryness. Yield of product was 1.54 g. The product was methylated with diazomethane at 4°C, and after examination by gas (Fig. 11) and t.l.c. it was separated by preparative-scale gas chromatography as described previously (Jones, 1969*a*). The separation resulted in the following amounts of material being available for analysis (starting weight 1200 mg).

Carbon number:	
26.0	225 mg
24.4	12 mg
23.6	28 mg
22.4	1 mg
21.2	0.1 mg

In addition, approximately 100 mg of a material with a carbon number greater than 26 was collected. Each fraction was subjected as far as possible to the following procedures:
(1) Mass spectroscopy, using an AEI MS-9 high resolution mass spectrometer.
(2) N.m.r. (proton resonance) spectroscopy, using a Varian model HA-100 spectrometer.
(3) I.r. spectroscopy, using a Perkin Elmer Model 257 recording spectrophotometer.
(4) Micro-combustion analysis.

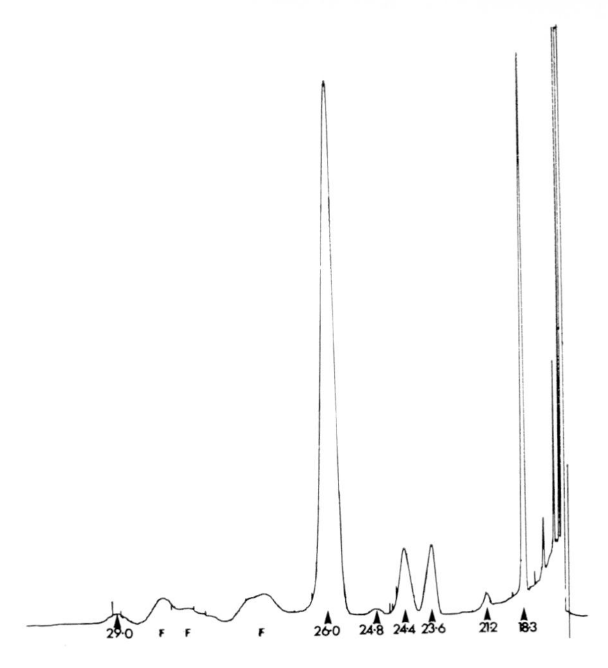

Figure 11. The actual reaction mixture separated by preparative scale gas chromatography (separated as methyl esters on FFAP (235°C) in a modified Pye model 104 chromatograph). Peaks labelled F are caused by polymers of formaldehyde.

Figures 12 & 13. I.r. absorbance and proton resonance spectra of methyl oleate and four of the isolated compounds. Each compound is identified by a code (TA, TC, TD, T.12) followed by three numbers, which are, respectively, its Carbon Numbers on FFAP, PEGA and APL, in that order.

The Carbon Nos. were determined on 5 ft. columns in a Pye model 124 chromatograph, with reference to palmitic acid (16:0) and stearic acid (18:0) as 16.0 and 18.0 respectively.

The i.r. spectra were recorded on a Perkin Elmer model 257 recording spectrophotometer; material was smeared on sodium chloride discs.

Nuclear magnetic resonance spectra were recorded on a Varian model HA.100 spectrometer at 100 megacycles. The material was in dilute solution (10–15%) in carbon tetrachloride containing tetramethyl-silane as reference.

Figure 12. I.r. absorption spectrum of methyl oleate. Note the C—H stretching vibration of the —CH=CH— group at 3000 cm^{-1}, and the group of absorptions characteristic of a —C—O— stretch in an ester linkage at 1240, 1195 and 1170 cms^{-1}.

In the i.r. spectrum of TA/26.0/26.5/21.8, absorptions at 1130, 1110 and 1080 cms^{-1} are assigned to the C—O stretching vibration in the group CH—O—CH_2OH. Absorptions at 1400 (weak) and 1040 (strong) are assigned to the C—O stretching vibration in the group —CH_2OH.

In the i.r. spectrum of TC/26.3/23.8/20.7, note the absence of an absorption at 3000 cms^{-1} (absence of CH=CH group). The ester carbonyl (C=O stretch) absorbance is at 1740 cms^{-1}, and three absorptions at 1240, 1190 and 1165 cms^{-1} are characteristic of —C—O— in the ester linkage.

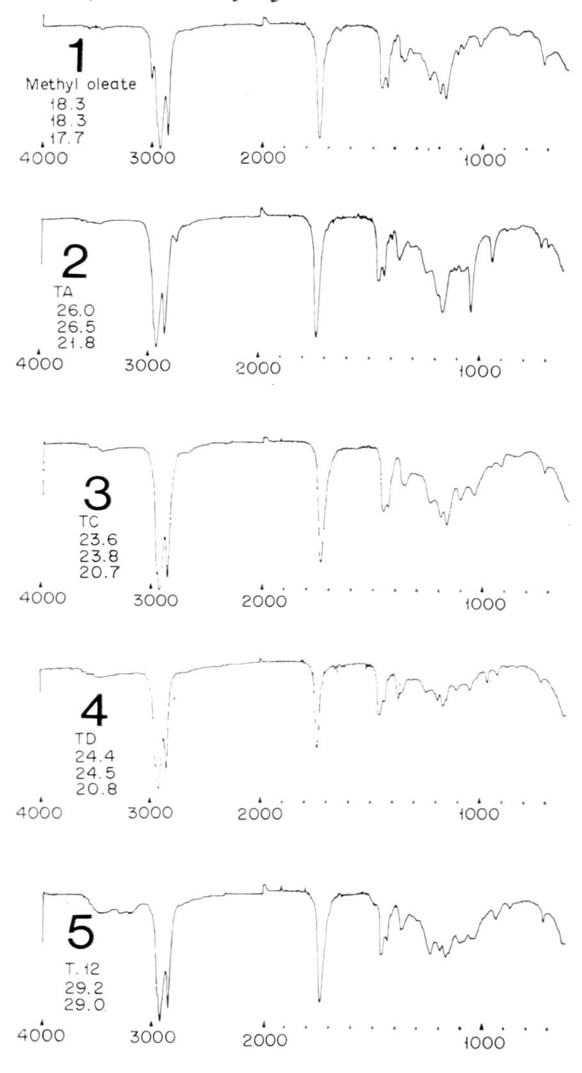

Figure 12

Absorptions at 1040 cms^{-1} and 1100 cms^{-1} are suggestive of a —CH_2OH and CHOH respectively. The i.r. spectrum of TD/24.4/24.5/20.8 is similar to that of compound TC/23.6/23.8/20.7, with the addition of an absorption at 965 cms^{-1}, tentatively assigned to —CH=CH— (trans) present in a contaminating compound.

There is a broad absorption in the i.r. spectrum of T.12/29.2/29.0/– at 3000–3500 cms^{-1} which could indicate hydrogen-bonded OH groups. There are also absorptions indicative of an ester (1240, 1195, 1170, 1150 cms^{-1}). A broad absorption at 1100 cms^{-1} suggests both CHOH and C—O—C groupings: an absorption at 1050 cms^{-1} suggests a —CH_2OH group.

I.r., proton resonance and mass spectra obtained from the separated adducts are shown in Figs. 12–19. A brief analysis of the more interesting points in each spectrum is included in the Figure caption, but the main analysis of the spectra is undertaken below; the figures refer to the *methyl esters* of the reaction products.

Compound TA (26.0/26.5/21.8)* was obtained in large yield and high purity from the reaction mixture separated. The structure, names and abbreviations assigned to the compound were:

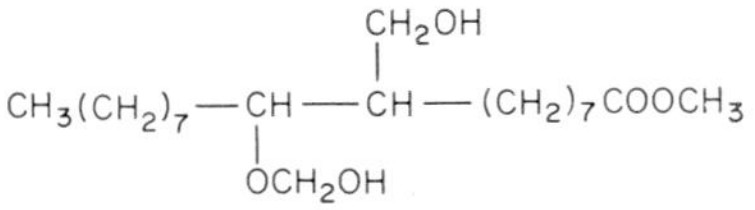

$$CH_3(CH_2)_7—\underset{\displaystyle OCH_2OH}{\underset{|}{CH}}—\overset{\displaystyle CH_2OH}{\overset{|}{CH}}—(CH_2)_7COOCH_3$$

Methyl 9-hydroxymethyl-10-hydroxymethyloxy-octadecanoate (10-OHHODA) and/or

* Reference titles refer to Table 10.

Figure 13. Proton resonance spectrum (100 mcs) of $\overset{18}{CH_3}—(CH_2)_5—\overset{12}{CH_2}—\overset{11}{CH_2}—\overset{10}{CH}=\overset{9}{CH}—\overset{8}{CH_2}—\overset{7}{CH_2}(CH_2)_4—\overset{2}{CH_2}\overset{1}{C}OOCH_3$, methyl oleate.

The triplet at $\delta = 5.2$ ppm is due to the protons of the —CH=CH— group, that at $\delta = 2.2$ ppm is due to protons of the α-methylene group. The sharp singlet at $\delta = 3.6$ ppm is caused by protons in the methyl group in the ester linkage. Protons in the allylic methylene groups (8, 11) are responsible for the signal at $\delta = 2.0$. Protons in the methylene groups at 7 and 12 are responsible for an amorphous signal at $\delta = 1.6$ ppm. The remaining methylene group protons give rise to a sharp signal at $\delta = 1.3$ ppm, and the terminal methyl group the signal at $\delta = 0.9$ ppm.

In the proton resonance spectrum of TA 26.0/26.5/21.8, two one-proton doublets (J = 6 c.p.s.) at $\delta = 4.45$ and 4.85 ppm are assigned to geminal non-equivalent protons in the $\rangle CH—O—CH_2—OH$ groups. The geminal non-equivalent protons of the $\rangle CH—CH_2OH$ are assigned to the signals at $\delta = 3.1$ ppm (triplet, J = 11 c.p.s. and 11 c.p.s.) and $\delta = 3.94$ ppm (quartet, J = 11 c.p.s. and 4.5 c.p.s.). The methine proton adjacent to oxygen gives a broad one-proton signal at $\delta = 3.25$ ppm.

There is a quartet signal at $\delta = 3.9$ ppm in the proton resonance spectrum of TC/23.6/23.8/20.7 and what appears to be a triplet at $\delta = 3.1$ ppm. These could be assigned to the two non-equivalent protons in the $\rangle$*CH—CH_2OH group (asymmetric carbon causing non equivalency). A signal at about $\delta = 3.3$ ppm is probably due to the methine proton in a $\rangle CH—OH$ group. The triplet at about $\delta = 1.15$ ppm is probably due to protons in one of the two methylene groups arrowed.

$$\overset{\downarrow}{—CH_2}—\underset{\displaystyle OH}{\underset{|}{CH}}—\underset{\displaystyle CH_2OH}{\underset{|}{CH}}—\overset{\downarrow}{CH_2}$$

The proton resonance spectrum of TD/24.4/24.5/20.8 is similar to that of compound TC/23.6/23.8/20.7, with the addition of weak signals at $\delta = 2.0$ ppm and 5.2 ppm, tentatively assigned to the presence of an unsaturated contaminant.

The triplet in the proton resonance spectrum of T.12/29.2/29.0/– due to the α methylene group has shifted downfield to $\delta = 2.3$ ppm. There is a one-proton triplet at $\delta = 2.95$, probably due to a non-equivalent proton in an oxymethylene bridge

$$(\underset{|}{\overset{|}{CH}}—CH_2—O—\underset{|}{\overset{|}{CH}})$$

the other proton being responsible for the quartet at about $\delta = 3.9$ ppm.

$$\begin{array}{c} OCH_2OH \\ | \\ CH_3(CH_2)_7—CH—CH—(CH_2)_7COOCH_3 \\ | \\ CH_2OH \end{array}$$

Methyl 9-hydroxymethyloxy-10-hydroxymethyl-octadecanoate (9-OHHODA). Carbons 9 and 10 are both asymmetric centres. This accounts for the non-equivalency of the methylene protons in both —CH_2OH groups, seen in the proton resonance spectrum (Fig. 13). The coupling constants measured (Jones & Williams, 1969) confirm the presence of coupled non-equivalent methylene protons. The i.r. spectrum (Fig. 12) shows absorptions which are also highly indicative of this type of structure. The mass

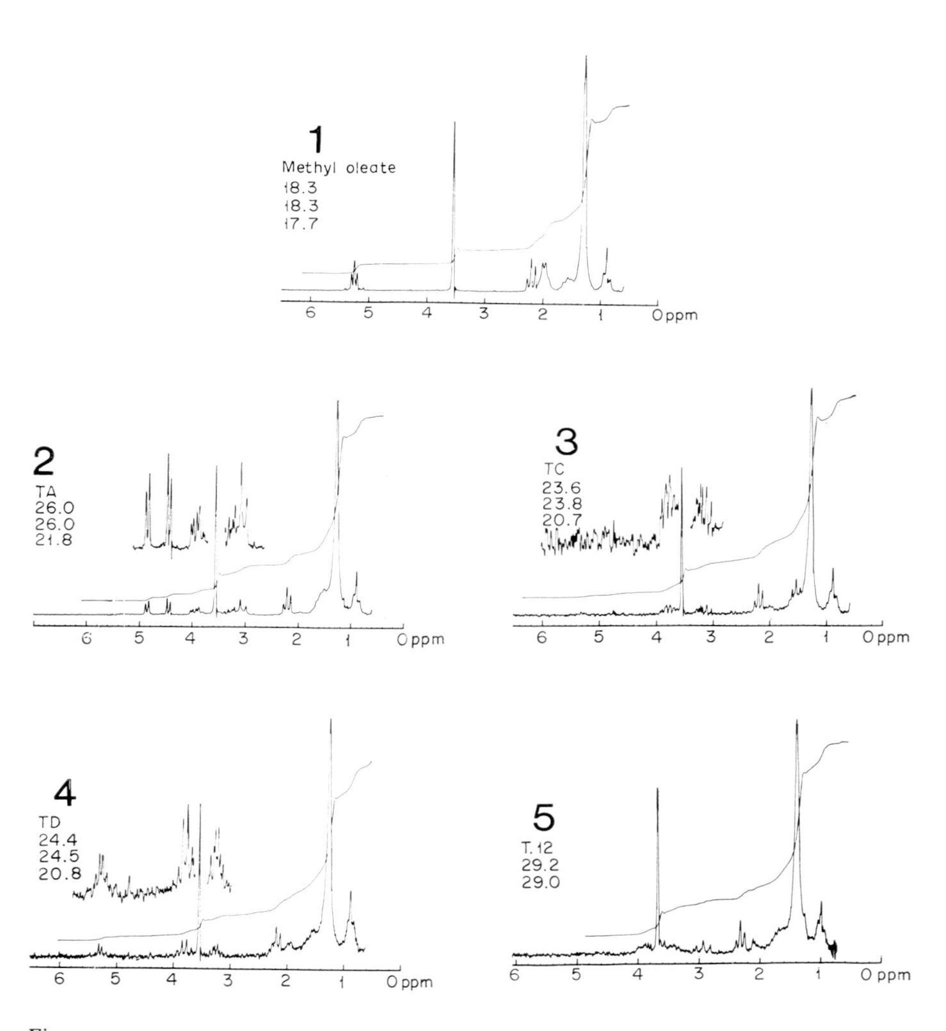

Figure 13

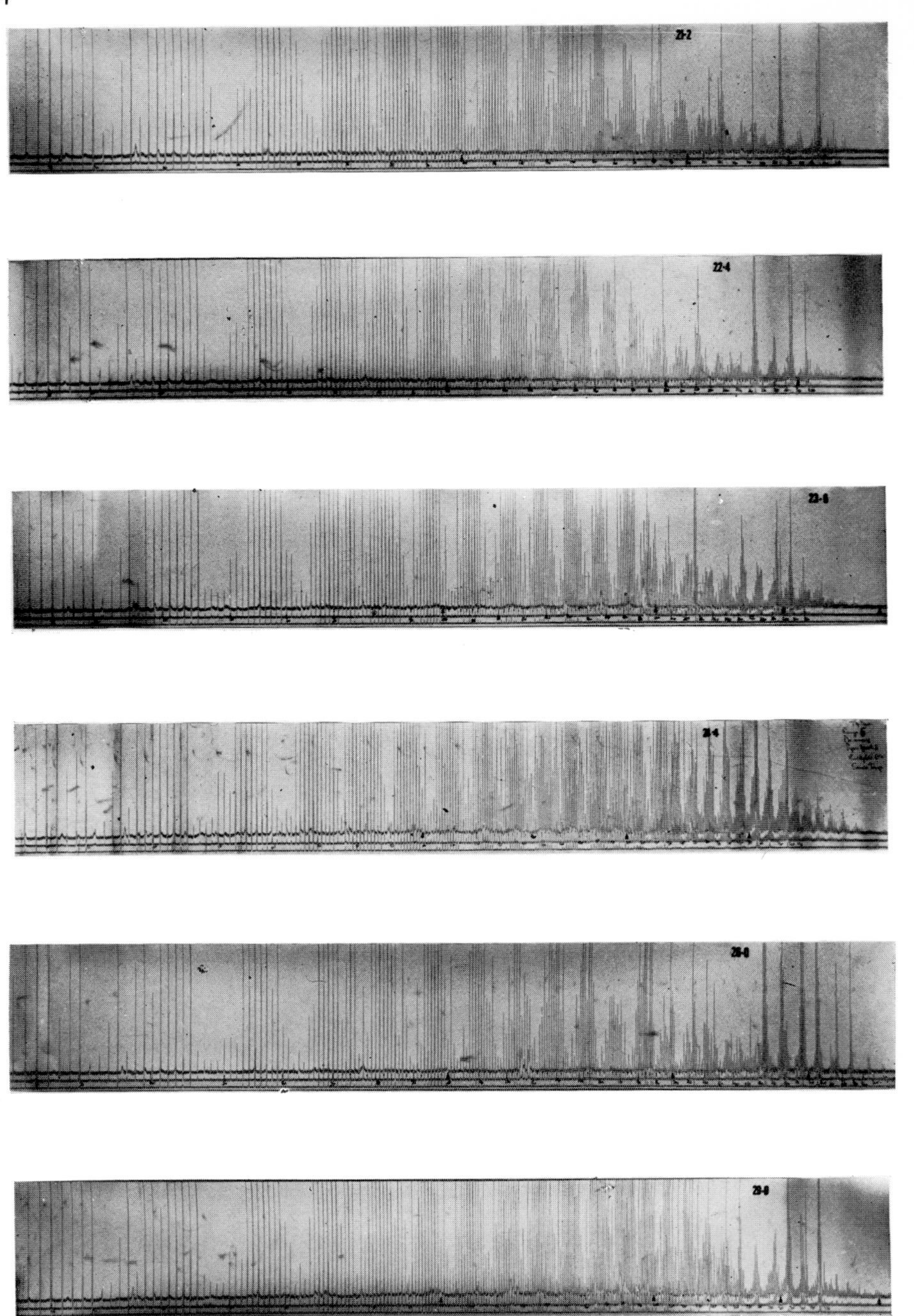

Figures 14–19. The mass spectra of compounds 21.2, 22.4, 23.6, 24.4, 26.0 and 29.0 (Carbon Nos. on FFAP). Recorded on an AEI MS-9 mass spectrometer.

spectrum (Fig. 18) does not show a molecular ion (M^+) peak, but this is a common occurrence with hydroxy compounds and alcohols. Peaks occurred at 357 (M–17) 343 (M-31) 327 (M-47), corresponding respectively to the loss of OH, CH_2OH, and OCH_2OH, 308 and 294. A microcombustion analysis performed immediately after separation gave C, 67.49%; H, 10.76%. Theoretical values for $C_{21}H_{42}O_5$ are C, 67.38%; H, 11.23%; a repeat microanalysis performed six months later suggested that the compound had lost formaldehyde in the intervening period, and this may be taken as confirmatory evidence for the presence of a hydroxymethyloxy group.

Compounds TC (23.6/23.8/20.7)* and TD (24.4/24.5/20.8)* were considered to be isomers, the slight difference in the proton resonance spectra (Fig. 13) being due to contamination of TD with an unsaturated compound (24.8/24.9/19.2)*.

The structure, names and abbreviations assigned are:

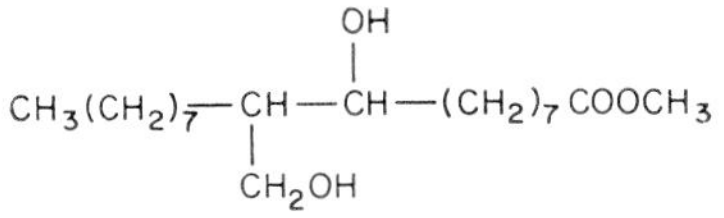

$$CH_3(CH_2)_7-\underset{\displaystyle CH_2OH}{\underset{|}{CH}}-\overset{\displaystyle OH}{\overset{|}{CH}}-(CH_2)_7COOCH_3$$

TC (23.6/23.8/20.7)*
methyl 9-hydroxy-10-hydroxymethyl-octadecanoate (10-HHODA).
TD (24.4/24.5/20.8)*

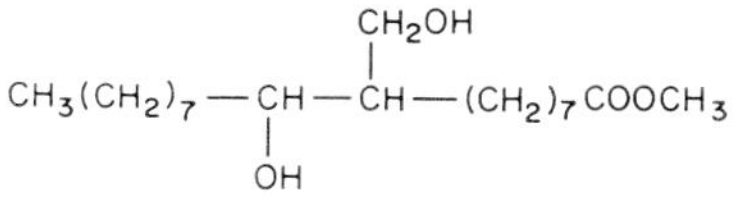

$$CH_3(CH_2)_7-\underset{\displaystyle OH}{\underset{|}{CH}}-\overset{\displaystyle CH_2OH}{\overset{|}{CH}}-(CH_2)_7COOCH_3$$

methyl 9-hydroxymethyl-10-hydroxy-octadecanoate (9-HHODA).

The cracking patterns observed in the mass spectra (Figs. 16, 17) are nearly identical for these two compounds; this could only mean that the compounds are isomers, or that they are rapidly interconvertible. The i.r. spectra (Fig. 12) are similarly nearly identical, the only difference being assigned to traces of (24.8/24.9/19.2)* in compound TD. The proton resonance spectra confirm the presence of a $-CH_2OH$ group, and a methine proton adjacent to oxygen, in each case.

Compound (24.8/24.9/19.2)* may be considered as a mixture of dehydration products:

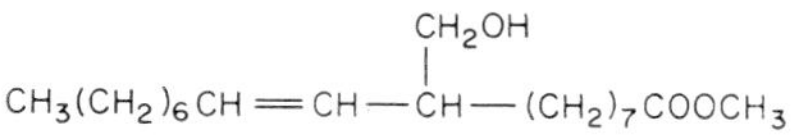

$$CH_3(CH_2)_6CH=CH-\overset{\displaystyle CH_2OH}{\overset{|}{CH}}-(CH_2)_7COOCH_3$$

methyl-9-hydroxymethyl-octadec-10-enoate (9-MHO) or

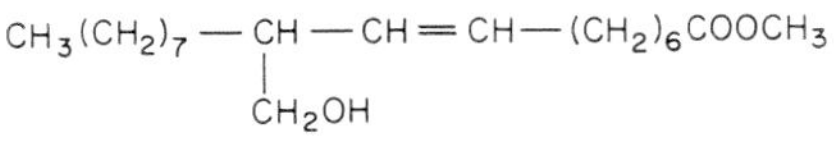

$$CH_3(CH_2)_7-\underset{\displaystyle CH_2OH}{\underset{|}{CH}}-CH=CH-(CH_2)_6COOCH_3$$

methyl-10-hydroxymethyl-octadec-8-enoate (10-MHO), or both. They are removed by

* Reference titles refer to Table 10.

hydrogenation, being converted into a compound with a carbon number of 24.4 on FFAP.

Blomquist *et al.* (1957*a*,*b*), in their studies of the high temperature reaction of formaldehyde and alkenes, suggested a single-stage reaction proceeding through a planar cyclic intermediate for the genesis of similar compounds. This seems a much more likely explanation for the presence of 9- and 10-MHO than the suggestion of proton-catalysed dehydration made by Artun (1966); if the latter occurred, then these compounds would be much more abundant in the reaction mixture than they are.

Compound T.3 (22.4/22.7/18.8)* could not be characterized so completely since little material (1 mg) was available for analysis. Examination of the cracking pattern (Fig. 15), and the rather poor i.r. spectrum, suggest that this compound is a 1,3-dioxan and the structures, names and abbreviations assigned are:

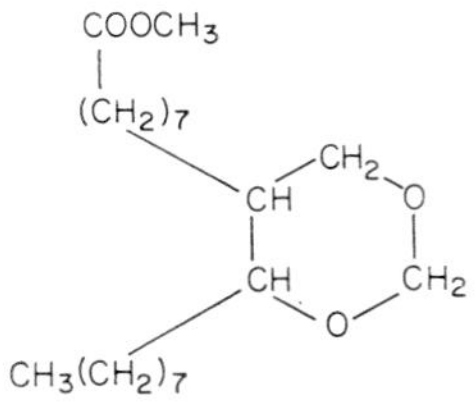

$COOCH_3$
$(CH_2)_7$
CH — CH_2 — O
CH — O — CH_2
$CH_3(CH_2)_7$

5(8-carbomethoxy-octyl)-6-octyl-1,3-dioxan(5-COOD)

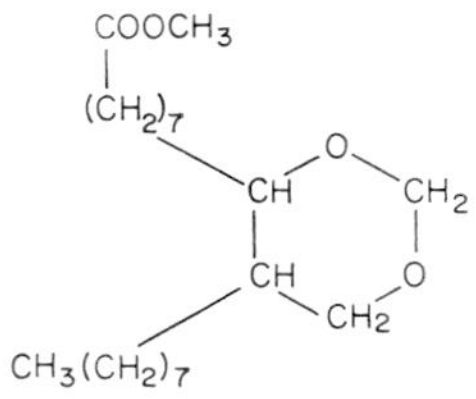

$COOCH_3$
$(CH_2)_7$
CH — O — CH_2
CH — CH_2 — O
$CH_3(CH_2)_7$

5-octyl-6(8-carbomethoxy-octyl)-1,3-dioxan(6-COOD)

These structures are in keeping with postulated mechanisms for the reaction.

Compound T.12 (29.2/29.0/—)* remains a mystery. The evidence (Figs. 12, 13, 19) suggest a high mol. wt. polar compound such as:

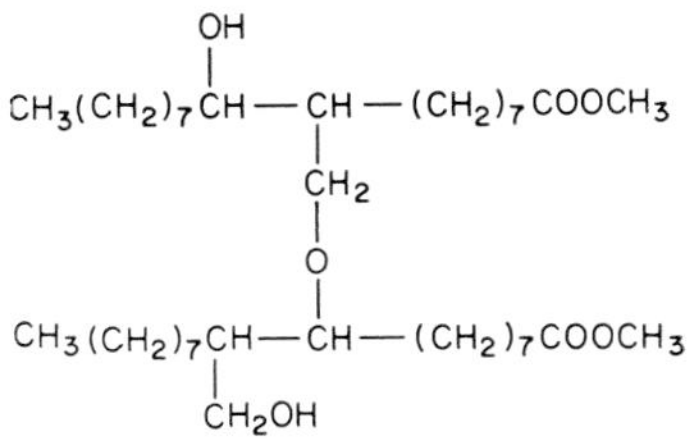

OH
$CH_3(CH_2)_7CH—CH—(CH_2)_7COOCH_3$
CH_2
O
$CH_3(CH_2)_7CH—CH—(CH_2)_7COOCH_3$
CH_2OH

If this is the case, this compound represents a vital stage in the conversion of fatty acids to a macromolecular network, and as such is an essential intermediate for the true 'fixation' of unsaturated fats.

Compound T2(21.2/21.2/18.3)* was converted to a compound of carbon number 21.0 (FFAP) by hydrogenation (see Figs. 20, 21 and 22) and, therefore, probably possesses a double bond. The mass spectrum (Fig. 14) indicates only that this compound resembles the other adducts (m/e of 308 and 294 are common to all the mass spectra). Tentative suggestions as to its structure are:

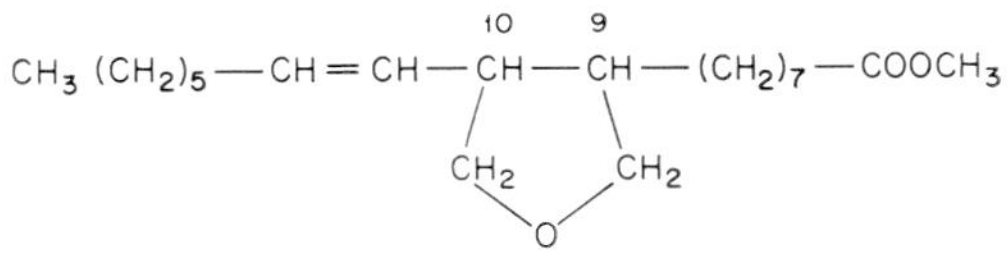

$$CH_3(CH_2)_5—CH=CH—\overset{10}{C}H—\overset{9}{C}H—(CH_2)_7—COOCH_3$$
$$\text{(C-10 and C-9 bridged by } CH_2—O—CH_2\text{)}$$

and

$$CH_3(CH_2)_7—\overset{10}{C}H—\overset{9}{C}H—CH=CH—(CH_2)_5COOCH_3$$
$$\text{(C-10 and C-9 bridged by } CH_2—O—CH_2\text{)}$$

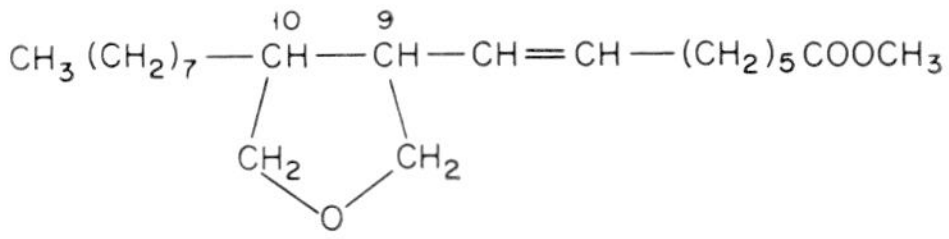

though compounds such as:

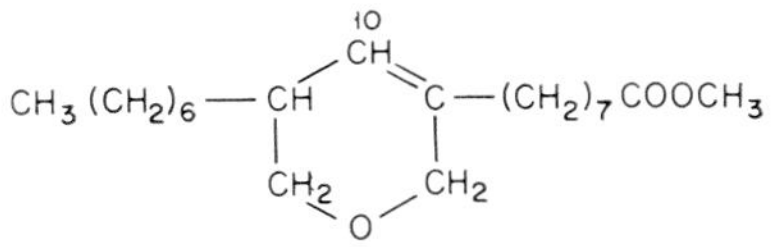

$$CH_3(CH_2)_6—CH—\overset{10}{C}H=C—(CH_2)_7COOCH_3$$
$$\text{(CH and C bridged by } CH_2—O—CH_2\text{)}$$

cannot be excluded.

The compound must have a 5- or 6-membered ring and no polar —OH groups to explain its short elution time, particularly on the APL columns.

(e) *Derivation of equilibrium constants and heat of reaction for the reaction of oleic acid with formaldehyde.* Batches of pure oleic acid (20 mg) were 'fixed' in 10 ml portions of 10% formalin-saline (pH 4.05). At various times five replicate samples were removed, extracted and methylated with diazomethane, and analysed on FFAP columns in a Pye model 124 chromatograph. The results, presented in Table 10 and graphically in Fig. 23, were subjected to a complete statistical analysis. They suggest that in the case of oleic acid, at room temperature, equilibrium is reached when about 96–97% of the original oleic acid is left (see Fig. 20). This is in contrast to the reaction occurring at 115°C, which proceeds substantially to completion in about 20 hr (equilibrium in this case representing an 88% change in the oleic acid). It is possible to deduce equilibrium constants and heats of reaction from the figures available; having made the customary assumptions and simplified the reaction to the 'pseudo' unimolecular:

$$\text{Oleic acid} \underset{k_2}{\overset{k_1}{\rightleftharpoons}} \text{Adduct}$$

the equilibrium constant $K(=k_1/k_2)$ was determined as

$$dc/dt = k_1(C_0—c)—k_2c$$

where dc/dt is the net forward rate of reaction, C_0 is the initial concentration of oleic

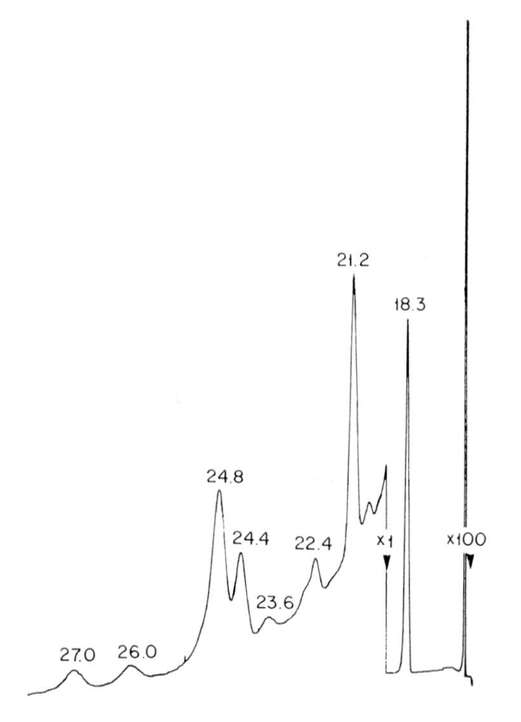

Figure 20. Oleic acid, fixed in 10% formalin-saline (pH 4.15) for 36 days. (Methyl esters separated on an FFAP column (255°C) in a Pye model 124 chromatograph.)

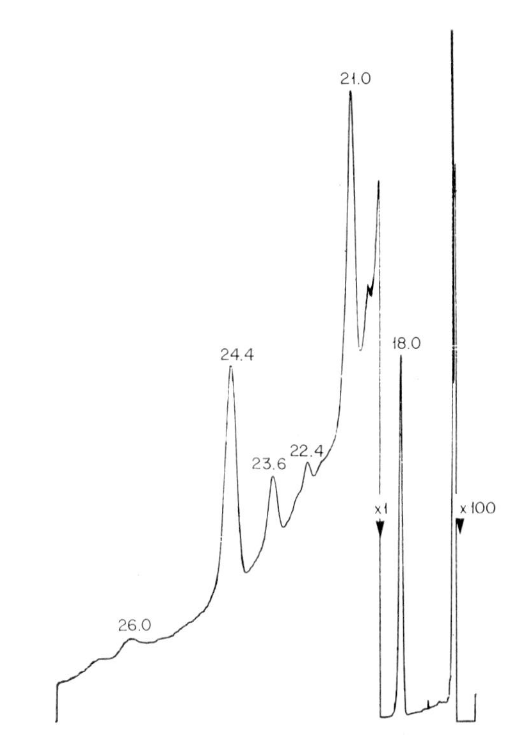

Figure 21. Oleic acid, fixed in 10% formalin-saline (pH 4.15) for 36 days, then hydrogenated (methyl esters separated on an FFAP column (255°C) in a Pye model 124 chromatograph).

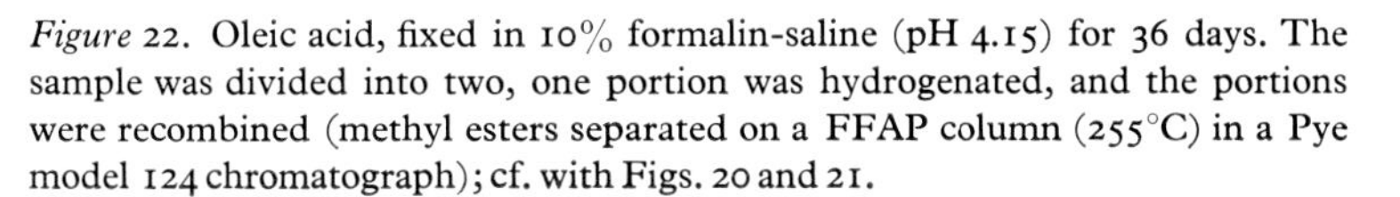

Figure 22. Oleic acid, fixed in 10% formalin-saline (pH 4.15) for 36 days. The sample was divided into two, one portion was hydrogenated, and the portions were recombined (methyl esters separated on a FFAP column (255°C) in a Pye model 124 chromatograph); cf. with Figs. 20 and 21.

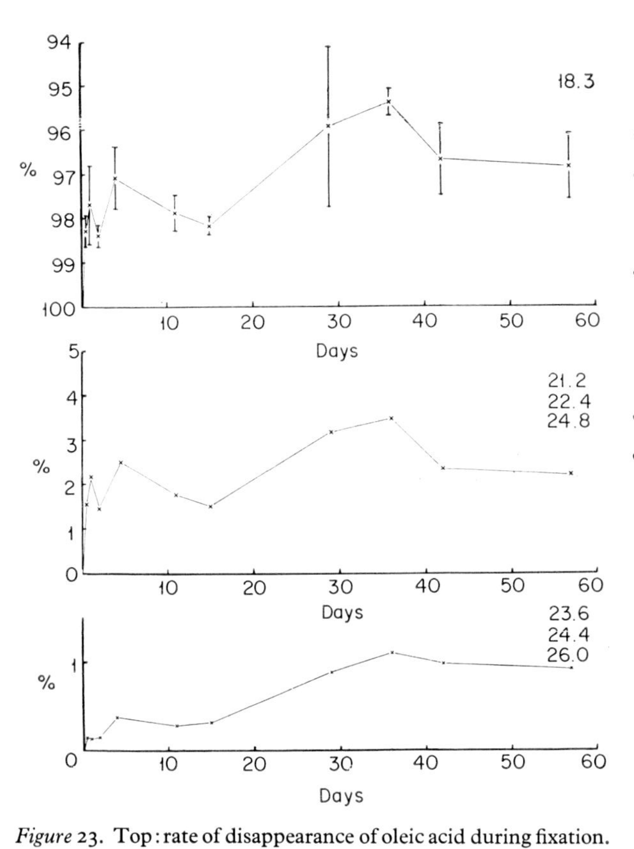

Figure 23. Top: rate of disappearance of oleic acid during fixation. Mean and standard deviation.
Middle: rate of accumulation of the 'end-products' of fixation.
Bottom: accumulation of 'intermediate products' of fixation.

Table 10. % Composition of 'oleic acid' after fixation in 10% formalin-saline for varying periods of time. Composition of the mixtures determined by chromatography of the methyl esters in a Pye model 124 chromatograph.

Time (days)	*Approximate Carbon Numbers on FFAP/PEGA/APL* *18.3/18.3/17.7	21.2/21.2/18.3	22.4/22.7/18.8	23.6/23.8/20.7	24.4/24.5/20.8	24.8/24.9/19.2	26.0/26.5/21.8
1 (0.5)	98.29±0.34	1.30±0.27	0.11±0.55	0.03±0.008	0.04±0.005	0.14±0.03	0.07±0.08
2 (1)	97.71±0.89	1.87±0.83	0.12±0.09	0.03±0.004	0.04±0.011	0.18±0.03	0.07±0.02
3 (2)	98.43±0.24	1.01±0.19	0.17±0.05	0.03±0.004	0.06±0.011	0.24±0.03	0.06±0.015
4 (4)	97.11±0.69	2.05±0.69	0.14±0.09	0.21±0.04	0.08±0.02	0.03±0.08	0.08±0.04
5 (11)	97.94±0.39	1.10±0.25	0.34±0.12	0.12±0.07	0.11±0.06	0.34±0.04	0.05±0.04
6 (15)	98.19±0.20	1.08±0.28	0.12±0.09	0.10±0.02	0.15±0.02	0.25±0.14	0.07±0.03
7 (29)	95.95±1.77	1.09±0.32	0.58±0.39	0.15±0.12	0.66±0.26	1.49±0.71	0.07±0.03
8 (36)	95.43±0.33	1.47±0.05	0.26±0.02	0.10±0.05	0.79±0.09	1.74±0.20	0.20±0.034
9 (42)	96.69±0.82	0.85±0.36	0.29±0.15	0.14±0.05	0.78±0.08	1.19±0.65	0.06±0.01
10 (57)	96.87±0.68	0.63±0.31	0.29±0.09	0.22±0.02	0.59±0.05	1.29±0.24	0.10±0.04

Notes: *18.3 represents all the fatty acids present in the original starting material; this was about 99% pure, containing a little $C_{16:1}$ and $C_{20:3}$. For purposes of calculation, the original starting material was taken as 100%.

The values for each measurement above are given as Mean ± S.D. for a group of five separate estimations.

Each column in the table has been subjected to analysis of variance (*F* test) and to Students *t* test between groups. The difference between the groups in each column was significant at the 0.1% level on the *F* test. Results for the *t*-tests are summarized in Table 11, each time being given a code between 1 and 10 (see left-hand side of table) to simplify presentation.

Table 11. Results of *t*-Tests.

Carbon No. of compound	*Significance at a level of:* 0.1%	1.0%	5.0%
18.3	1, 2, 3, 5, 6 × 7, 8	1, 3, 5 × 9, 10	1, 3 × 4
18.3		4 × 8	1, 8 × 9
17.7		6 × 9	4 × 6, 7
		8 × 10	6 × 10
21.2	2 × 9, 10	1 × 4	1 × 2, 10
21.2	3 × 4	2 × 3, 5, 6, 7	4 × 8
18.3	4 × 5, 6, 7, 9, 10	8 × 10	8 × 9
22.4	1, 2, 3, 4, 6 × 7	7 × 8, 9, 10	1, 2, 4 × 5
22.7			5 × 6, 7
18.8			
23.6	1, 2, 3 × 10	1 × 9	1 × 5, 6, 8
23.8	6, 8 × 10	2, 3 × 5, 9	2, 3 × 6, 8
20.7		4 × 6, 8	4 × 5, 9
		5 × 10	7, 9 × 10
24.4	1, 2, 3, 4, 5, 6 × 7, 8, 9, 10	8, 9 × 10	7 × 8, 9
24.5			
20.8			
24.8	1, 2, 3, 4, 5, 6 × 7, 8, 9, 10	—	8 × 9, 10
24.9			
19.2			
26.0	1, 2, 3, 4, 5, 6, 7 × 8	—	5 × 10
26.5	8 × 9, 10		
21.8			

acid, and c is the amount of oleic acid reacted, after time t. These may be directly replaced by percentages, taking C_0 as 100%, and c as 4% and 88% for the low- and high-temperature reactions respectively.
At equilibrium, $dc/dt = 0$, therefore

$$k_1(C_0 - c) = k_2 c \text{ or } k_1/k_2 = c/C_0 - c$$

Substituting the percentage values for C_0 and c,

$$K_{22^\circ} = 4/96 \text{ and } K_{115^\circ} = 88/12 = 7.34$$

The heat of reaction was derived from the formula:

$$\log (K_2/K_1) = H/eR\,(1\,T_2 - 1\,T_1) = H\,(T_2 - T_1)/4.576 T_1 T_2$$

$$T_1 = 295^\circ A,\ T_2 = 388^\circ A$$

Using this formula the heat of reaction was calculated as +12.63 kcals. The reaction was, therefore, endothermic and the heat intake per gram molecule of oleic acid reacting was approximately 12.6 kcals. In view of the assumptions made in calculating the heat of reaction and equilibrium constant, these were considered as only approximate.

Discussion

Chemical aspects of the reactions

The schemes on the opposite page show the course of the reactions that occur between the double bond in oleic acid and formaldehyde, under conditions of normal fixation. The numbers in brackets are the Carbon Nos. of the methyl esters on FFAP columns, and refer particularly to Fig. 20.

Several minor constituents of reaction mixtures have at times been observed. (Carbon No. on FFAP 25.4, 25.6 etc.). These could correspond to possible intermediates for (21.2). The reaction scheme correlates well with other information derived; for example, carbon numbers on an APL column correspond approximately to the actual number of carbon atoms in the molecule. APL is only slightly sensitive to shape and polarity of molecules, particularly those, as in this case, derived from aliphatic hydrocarbons; therefore, the OHHODA compounds have a Carbon No. of 21.8, the HOODA compounds 20.7–29.8 .9 and 10—MHO have a double bond, and thus elute far more rapidly (19.2) than their saturated analogues (20.6). FFAP and PEGA are far more sensitive to shape and polarity than actual size of molecules. Thus on these materials the OHHODA's elute far more slowly than the COOD's, though both have the same number of carbon atoms.

The Prins reaction has previously been studied mainly from the stereochemical aspect. With cyclic alkenes, the reaction is highly stereospecific giving rise slowly to *trans* addition across the double bond. Blomquist & Wolinsky (1957) and Dolby (1962, 1963) have explained this in terms of solvation by the glacial acetic acid used as solvent.

$^+CH_2OH$ +   H   O CH_2 → ^-OAC +   H   O CH_2 →   OAC CH_2OH

The acetate ion attacks from the rear, giving overall *trans* addition. A similar mechanism

'*Fixation*' *of oleic acid*

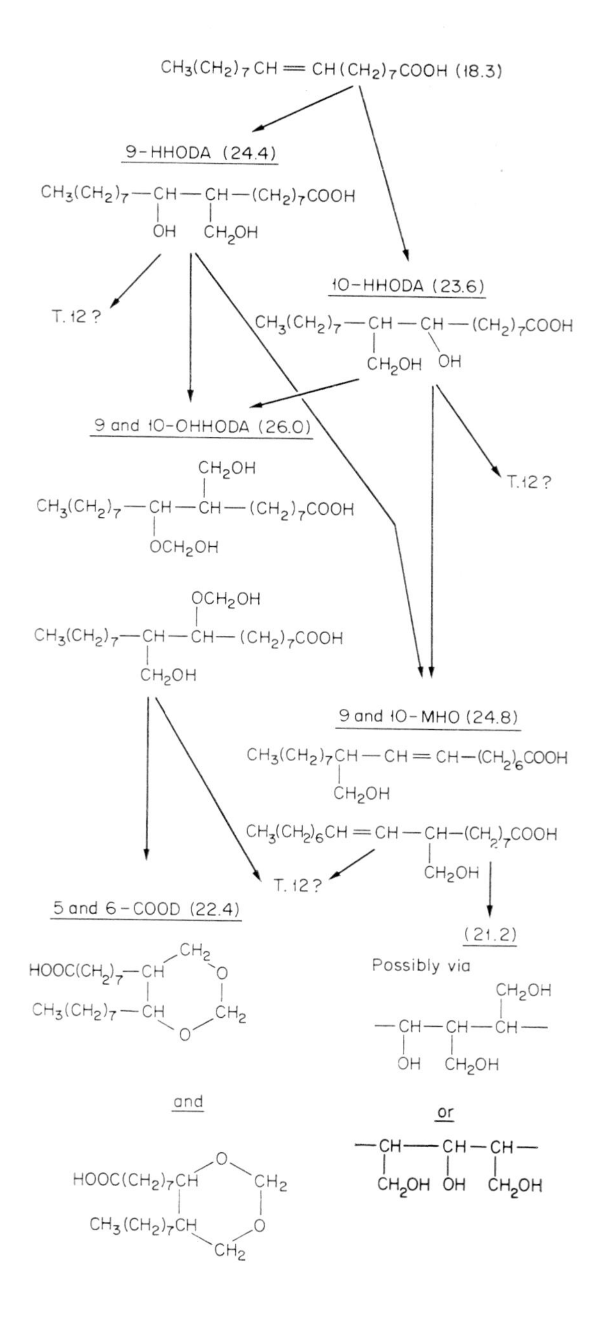

The fixation can be explained mechanistically by:

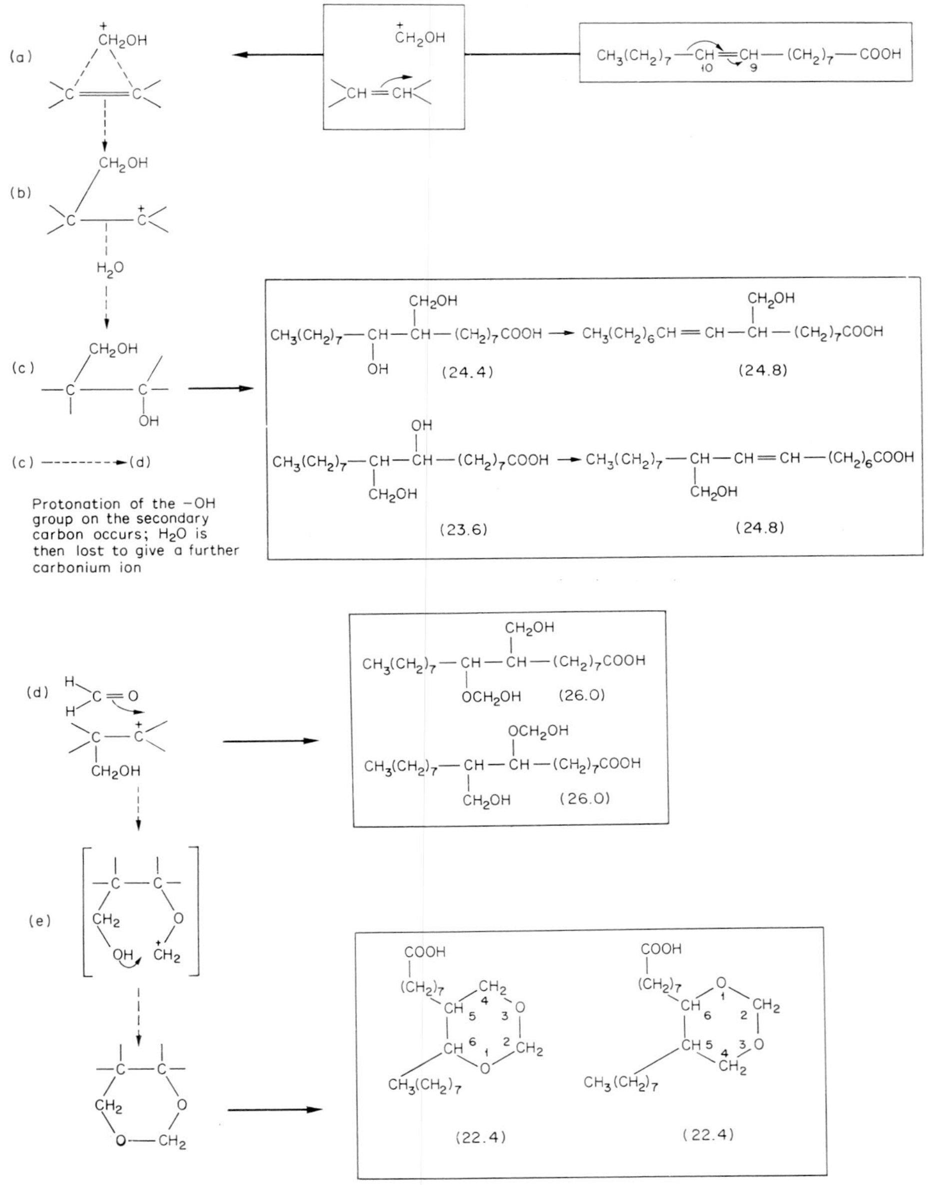

(a) $\overset{+}{C}H_2OH$ — $>C{=}C<$ ← $\overset{+}{C}H_2OH$, $>CH{=}CH<$ — $CH_3(CH_2)_7{-}CH{=}CH{-}(CH_2)_7{-}COOH$ (10, 9)

(b) CH_2OH, $>C{-}\overset{+}{C}<$

H_2O

(c) CH_2OH, $-C-C(OH)-$ →

$CH_3(CH_2)_7{-}CH(OH){-}CH(CH_2OH){-}(CH_2)_7COOH$ (24.4) → $CH_3(CH_2)_6CH{=}CH{-}CH(CH_2OH){-}(CH_2)_7COOH$ (24.8)

$CH_3(CH_2)_7{-}CH(CH_2OH){-}CH(OH){-}(CH_2)_7COOH$ (23.6) → $CH_3(CH_2)_7{-}CH(CH_2OH){-}CH{=}CH{-}(CH_2)_6COOH$ (24.8)

(c) ---------→ (d)

Protonation of the −OH group on the secondary carbon occurs; H_2O is then lost to give a further carbonium ion

(d) $H_2C{=}O$, $>C(CH_2OH){-}\overset{+}{C}<$ →

$CH_3(CH_2)_7{-}CH(OCH_2OH){-}CH(CH_2OH){-}(CH_2)_7COOH$ (26.0)

$CH_3(CH_2)_7{-}CH(CH_2OH){-}CH(OCH_2OH){-}(CH_2)_7COOH$ (26.0)

(e) [$-C-C-$, CH_2, O, OH, $\overset{+}{C}H_2$]

$-C-C-$, CH_2, O, O—CH_2 →

COOH, $(CH_2)_7$, CH 5, CH_2 4, O 3, CH_2 2, O 1, CH 6, $CH_3(CH_2)_7$ (22.4)

COOH, $(CH_2)_7$, CH 6, O 1, CH_2 2, O 3, CH_2 4, CH 5, $CH_3(CH_2)_7$ (22.4)

could be invoked to include attack from the rear by ^{-}OH in the case of the oleic acid reaction.

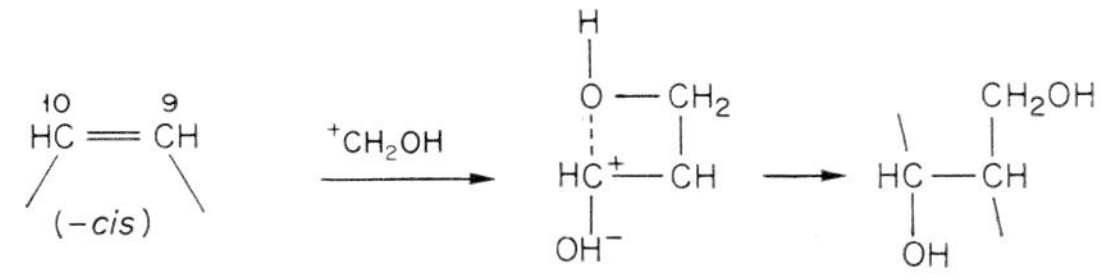

H
O — CH_2
10 9
HC ═ CH (-*cis*) $\xrightarrow{^{+}CH_2OH}$ HC^{+} — CH → HC — CH
OH^{-} OH CH_2OH

OH^{-} is however much smaller than CH_3COO^{-}, and an attack from in front would not be completely barred. In the case of oleic acid there is no means of checking the ratio of *trans:cis* addition as the single bond formed between carbons 9 and 10 permits free rotation to attain the configuration with minimum energy content. The results obtained with elaidic acid 'fixed' in formaldehyde were inconclusive; the differing physical states of the compounds at room temperature effectively precluded the possibility of distinguishing a difference in reactivity between oleic acid (18:1 *cis* isomer-liquid) and elaidic acid (18:1 *trans* isomer-solid) during fixation.

In the only experiment performed with elaidic acid under forcing conditions, it behaved exactly like oleic acid. Since there is free rotation round the single bond formed in the first stage of the reaction of oleic acid with formaldehyde, and there is no hindrance to attack in either compound, no differences in reactivity were in fact expected.

As far as the Prins reaction with isolated double bonds is concerned, other authors have postulated widely different mechanistic pathways; Zimmerman & English (1953) and Le Bel *et al.* (1963) favour a solvated trimethylene oxide as an intermediate; using their mechanisms, the reaction with oleic acid becomes:

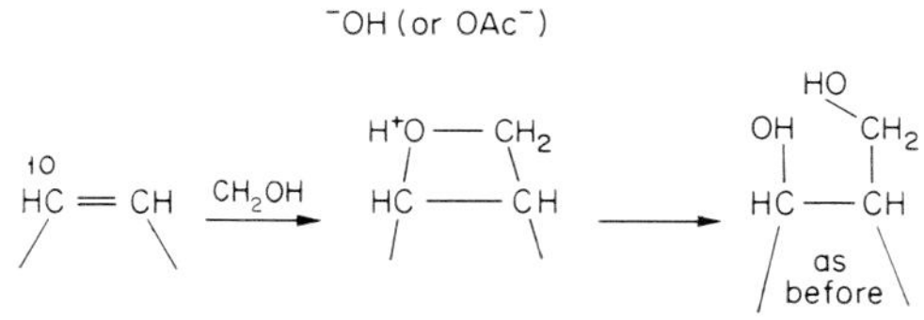

^{-}OH (or OAc^{-})
HO
$H^{+}O$ — CH_2 OH CH_2
10
HC ═ CH $\xrightarrow{CH_2OH}$ HC — CH → HC — CH
as before

If, instead of breaking the bond between C_{10} and the oxygen with an ^{-}OH, the oxygen is 'solvated' with a CH_2OH group the following is obtained:

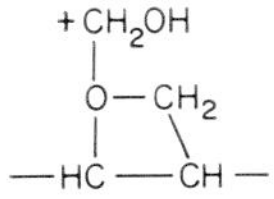

$+CH_2OH$
O — CH_2
— HC — CH —

An ^{-}OH can now be used to break the other carbon-oxygen bond in the tri-methylene oxide residue, giving the compound:

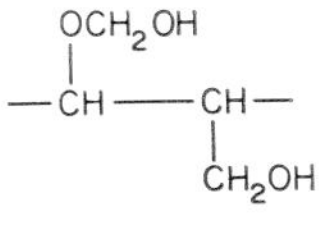

OCH_2OH
— CH — CH —
CH_2OH

The mechanism of Le Bel *et al.* does not explain the stereo-selectivity in the reaction between formaldehyde and cyclohexene, but it is the only mechanism that can be used to explain the occurrence of the OHHODA compounds; no previous analysis of the

Prins reaction has suggested this type of compound as a major product or even as a product at all. Dolby (1962), investigating the reaction of cyclohexene and formaldehyde, isolated compounds of the structures

CH_2OH OH IV · HO CH_2 O CH_2 V · CH_2OH VI

O O VII · CH_2 O CH_2 OH VIII

and his suggested mechanisms for the synthesis of these compounds agree closely with the sequence of reactions postulated above for the reaction of formaldehyde and oleic acid. Thus compounds IV, VI and VII above have their analogues in my system, and V and VIII may have analogues among the trace components of the 'hot' reaction, barely visible in the gas chromatograph trace in Fig. 11. Dolby suggests that the 1,3-dioxan VII is a cyclic 'formal'. His use of this name implies the formation of a methylene bridge more or less instantaneously between the oxygen atoms of the 1,3-diol, whereas I suggest, that though this compound is a 'formal' in structure, in the oleic acid-formaldehyde system it is derived from a precursor compound of the structure:

—CH—CH— ; O (CH_2OH) on the first CH, CH_2OH on the second CH (e.g. OHHODA)

It is obviously difficult to define a clear boundary between my mechanistic pathway and the classical mechanisms for 'acetal' formation. However, the significant point is that Dolby isolated no precursor of this 'formal', and in my case the precursor forms the major part of the reaction mixture. Dolby admittedly used glacial acetic acid as a solvent and sulphuric acid as catalyst, but this in itself is not sufficient explanation, and it must be assumed that in the oleic acid-formaldehyde system the structure:

—CH—CH— ; OCH_2OH on the first CH, CH_2OH on the second CH (OHHODA)

is thermodynamically favoured, whereas it is not so in other systems so far investigated. It may, in these other systems, have such a short 'life' that the formation of the 1,3-dioxan *does* approximate to the classical mechanism of 'acetal' formation.

Numerous other workers, in addition to those mentioned in the introduction and above have investigated stereochemical or mechanistic aspects of the Prins reaction; particular mention may be made of Baker (1948) and his study of the formation of the

1,3-glycols. Baker points out that Lewis acids such as aluminium chloride can be used in the Prins reaction in place of the normal proton donating acid. The reactive species would in this case be

$$\overset{\delta+}{CH_2}=\overset{\delta-}{OAlCl_3}$$

Price (1946) contrasted the classical Prins reaction with the chloromethylation of e.g. benzene.

$$C_6H_6 + CH_2O + HCl \xrightleftharpoons{ZnCl_2} C_6H_5CH_2Cl + H_2O$$

He considers that this represents the Prins reaction in an aromatic system, i.e. that the mechanisms are in effect identical. The work of Artun (1966) is notable in that he has suggested a reaction with oleic acid, and has proved that this occurs. He has not, unfortunately, elucidated the structures of his reaction products, but his work may serve as a valuable confirmation of the work presented in this paper. He mentions in the description of his experimental procedure that heat was evolved in the reaction. This does not agree with my suggestion that the reaction is endothermic, but some protonation of the carboxyl group may have occurred using his method, and this would give rise to heat evolved.

In the oleic acid-formaldehyde system, positional isomers can be, and are, formed in both the 'hot' and the 'cold' reaction. This can be attributed to the very slight inductive effect present, which even at room temperature is incapable of causing more than a slight accentuation of the attack at Carbon 9:

$$CH_3-(CH_2)_7-\underset{10}{CH}=\overset{9}{CH}-(CH_2)_7-COOH$$

The arrows indicate inductive effects due to the methyl group and the carboxyl group. Both these effects are very slight, due to the saturated chain of methylene groups interposed. However, there is a possibility that the arrangement of the molecule could be such that the carboxyl group would be spatially near the double bond and could have a polarising effect.

These effects are, at the best, weak. At room temperature the $^{+}CH_2OH$ electrophile will, therefore, exhibit only a slight preference for attack at Carbon 9 so that the mixtures of compounds will occur. The slight energy barrier to addition at Carbon 10 will be completely overcome at higher temperatures. This can be seen in the formation of the 1,3-glycols (contrast Figs. 11 & 20: Carbon Nos. 23.6 and 24.4), for

$$9-\text{HHODA} \qquad -\underset{OH}{\overset{10}{CH}}-\underset{CH_2OH}{\overset{9}{CH}}-$$

predominates at low temperatures, but 9- and 10-HHODA occur in approximately equal parts in the high temperature reaction.

In a few experiments performed in low ambient temperatures, only '9 hydroxymethyl-10-hydroxy' addition was seen.

My previous observation (Jones, 1969*a*), that linoleic acid reacts much more readily than oleic acid with formaldehyde, is confirmed here. The same trend is seen in the reaction with acrolein. The reactivity of linoleic acid is more than double that of oleic acid, and to explain this fact it must be assumed that the doubly activated allylic methylene group in linoleic acid (arrowed below) plays a considerable role:

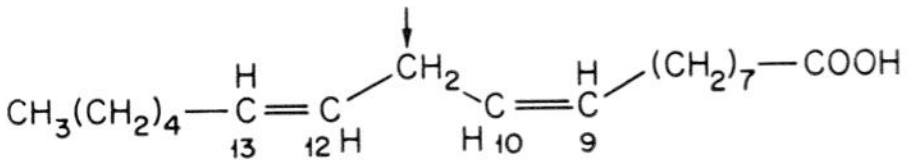

$$CH_3(CH_2)_4-\underset{13}{\overset{H}{C}}=\underset{12}{C}H-\overset{\downarrow}{C}H_2-\underset{10}{C}H=\underset{9}{\overset{H}{C}}-(CH_2)_7-COOH$$

i.e.

$$CH_3(CH_2)_4-\underset{H}{C}=\underset{H}{C}-\overset{H}{\underset{H}{C}}-CH=\underset{H}{C}-(CH_2)_7-COOH$$

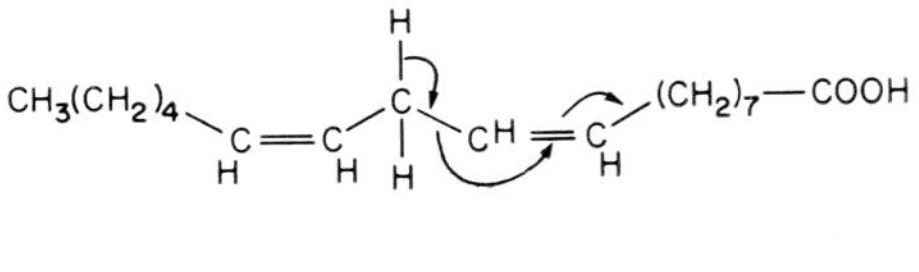

Under the reaction conditions used, free radical mechanisms must of necessity play no part, so we may assume that after initial attack of the carbonium ion (e.g. $^+CH_2OH$) we get bond migration with expulsion of an allylic proton and formation of a conjugated unsaturated system.

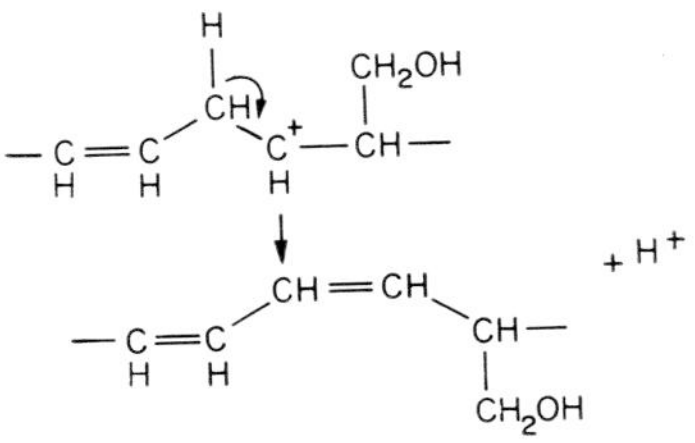

$$-\underset{H}{C}=\underset{H}{C}-\overset{H}{C}H-\underset{H}{\overset{+}{C}}-\overset{CH_2OH}{CH}-$$

$$\downarrow$$

$$-\underset{H}{C}=\underset{H}{C}-CH=CH-\underset{CH_2OH}{CH}- \quad + H^+$$

This is considerably more reactive, and explains satisfactorily the greatly increased reactivity of linoleic acid. Without further evidence, derived from the isolation and analysis of the reaction products, it is impossible to do more than theorize over the nature of the compounds formed by the reaction of acrolein with oleic acid. That the reaction is much more complex is evident from Fig. 3; under identical conditions, at least twelve different products are obtained from the reaction of acrolein with oleic acid while only four are obtained with formaldehyde. One factor contributing to this complexity is the fact that protonation of acrolein can give rise to two carbonium ions:

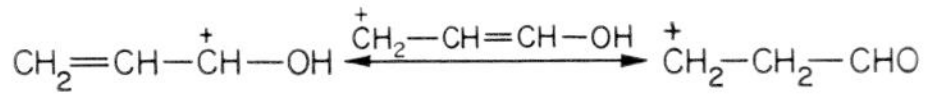

$$CH_2=CH-\overset{+}{C}H-OH \longleftarrow \overset{+}{C}H_2-CH=CH-OH \longrightarrow \overset{+}{C}H_2-CH_2-CHO$$

and, therefore, corresponding to the initial stage of the reaction of alkenes with formaldehyde

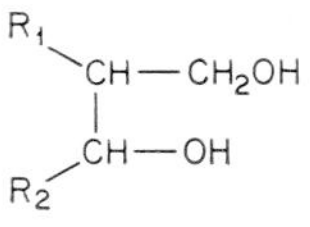

$$\begin{matrix} R_1 \\ & CH-CH_2OH \\ & | \\ & CH-OH \\ R_2 \end{matrix}$$

the following can be obtained:

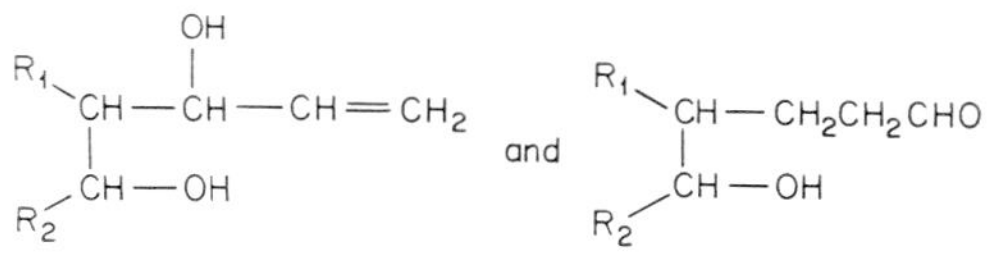

$$R_1-CH(-CH(OH)-CH{=}CH_2)-CH(OH)-R_2 \quad \text{and} \quad R_1-CH(-CH_2CH_2CHO)-CH(OH)-R_2$$

The possibilities of further reaction increase the complexity still further.

Biological aspects of the reactions

Since it now appears that lipids in tissues undergo a considerable number of reactions with aldehydes during fixation, the fact that it is difficult to demonstrate lipid histologically in tissues which have been stored in formaldehyde for more than four or five years (personal communications from G. A. Gresham, B. M. Herbertson and J. H. Rack) is not surprising. This is in part due to the conversion of the unsaturated fatty acids to 1,3-glycols, followed by irreversible conversion to other products; in view of the rather poor equilibrium constant at room temperature, this reaction might not appear to play much part in processes involving fatty acids in *the tissues* during routine histological fixation. That this view is erroneous is amply shown by Tables 8 and 9. The fatty acid patterns of both sebaceous adenoma and carcinoma are unrecognisable after two years fixation. Though for convenience the reaction has been considered an equilibrium reaction, it is reversible only within narrow limits. Thus it can be represented basically by

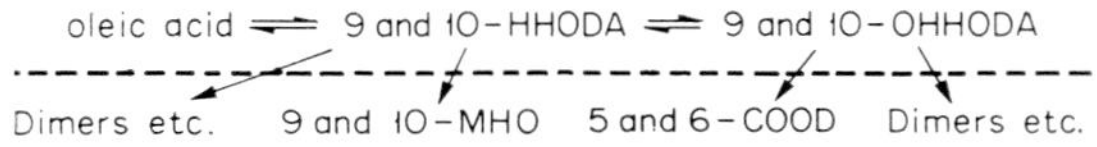

oleic acid ⇌ 9 and 10-HHODA ⇌ 9 and 10-OHHODA

↓ Dimers etc. ↓ 9 and 10-MHO ↓ 5 and 6-COOD ↓ Dimers etc.

The initial equilibrium is set up solely between oleic acid and the 1,3-glycols. This is a truly reversible reaction and is the subject of the equilibrium constant previously derived. The 1,3-glycols give rise to the other products, slowly, by irreversible reactions; being thus removed from the initial equilibrium, more oleic acid reacts, and eventually the reaction will approach completion. This is in agreement with the findings in Tables 8 and 9.

The fate of the 'other products' is important. Some may be true polymers, admittedly of a low mol. wt., but still unaffected by the emulsifying effect of phospholipids. It is likely that some will dissolve in the fixative; possession of one or two —OH groups, or even an ether linkage, will make fatty acid molecules decidedly more hydrophilic and the chance of losing them in the fixative greater.

The reaction of formaldehyde with double bonds during fixation occurs most readily when the double bond is present in a hydrophilic lipid (compare the effects of fixation on triolein and oleic acid). On theoretical grounds alone it may be assumed that the reaction would occur most readily with double bonds in hydrophilic lipids, such as free fatty acids and phospholipids, and it is significant that these species are the most labile during fixation, both showing a considerable flux from tissue to fixative (20–50%) during the first 100 days of fixation (Jones, unpublished observations). Hammar (1924), Kutschera-Aichbergen (1925, 1929), Kimmelstiel (1929) and Halliday (1939) have made similar observations, mainly on phospholipids, but attribute them to the 'altered nature' of the lipids after fixation, rendering them non-extractable by lipid solvents. Phospholipids do indeed react rather rapidly with formaldehyde, but some possible reactions, e.g. the conversion of $—NH_2$ to $N=CH_2$ in phosphatidyl ethanolamine and phosphatidyl serine with a concurrent loss of polarity, should render such altered molecules substantially *more* soluble in lipid solvents. Heslinga & Deierkauf (1961) suggested that phosphatidyl ethanolamine did in fact react rapidly with formaldehyde, but that other phosphatides reacted very slowly, if at all. At least three phospholipids (phosphatidyl choline or lecithin, phosphatidyl ethanolamine or cephalin, and sphingomyelin) have greatly increased solubility in formaldehyde solutions, only a minor part of which is due to hydrolysis (Jones, unpublished observations). Halliday (1939) had also observed this point and she noticed that lecithin, in solution in formalin, cannot be extracted with either alcohol-ether or chloroform. This latter point can be explained easily, for neither ether nor chloroform are good solvents for lecithin, which is in any case progressively more difficult to extract as the pH falls below 7. Whether this is due to a chemical reaction or to physical factors, is at present uncertain. Roozemond (1969*a*) found that formaldehyde inhibited the release of phospholipids from cryostat sections into water; he attri-

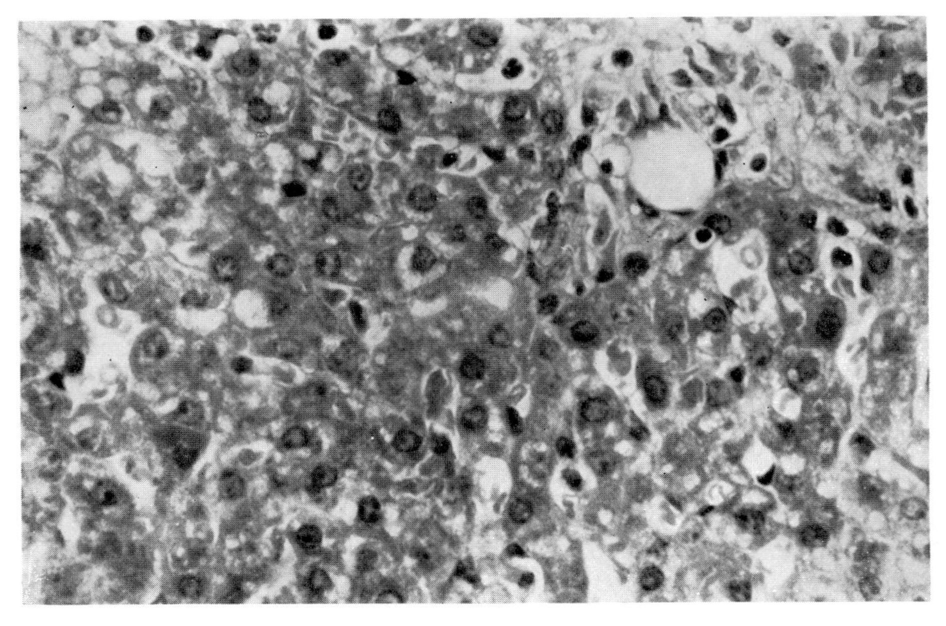

Figure 24

Figures 24–26. A comparison of three fixatives:
Fig. 24: formalin-saline
Fig. 25: buffered formalin
Fig. 26: acrolein-saline
In each case, the photomicrographs are of paraffin sections of ferret liver fixed in the appropriate fixative (Table 1), stained with Haematoxylin and Eosin under identical conditions. × 890

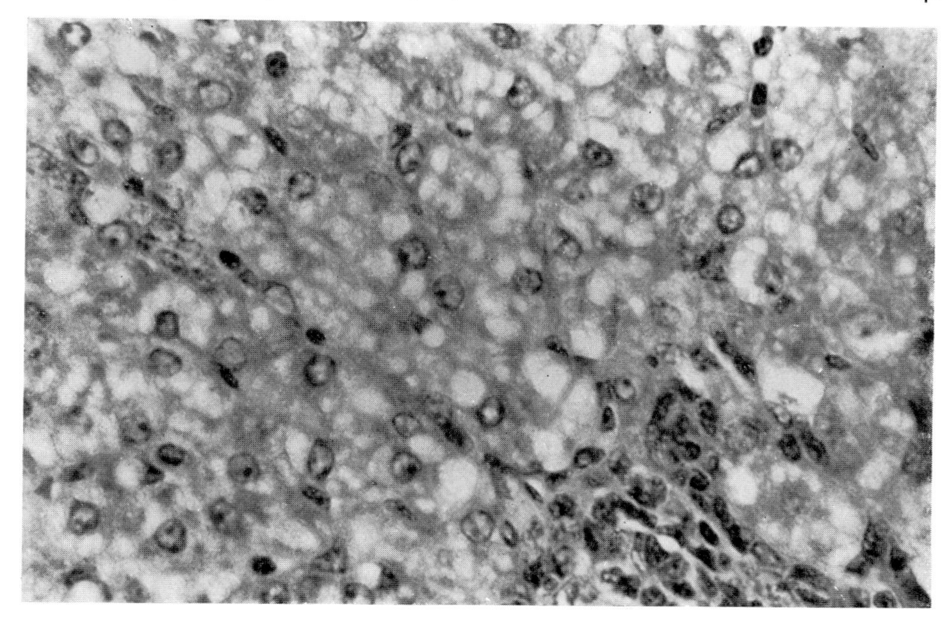

Figure 25

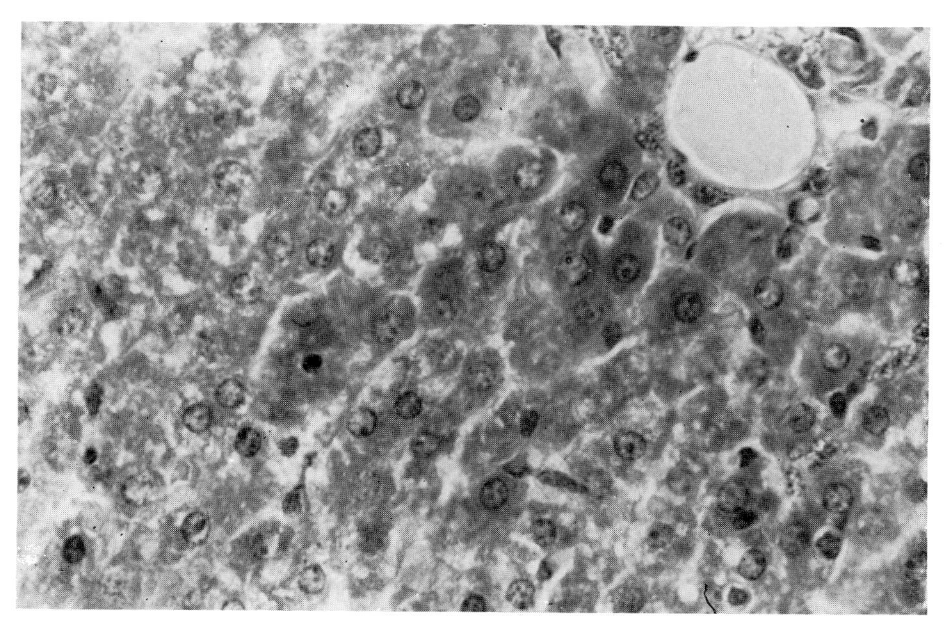

Figure 26

buted this finding to denaturation of protein by the formaldehyde, resulting in retention of phospholipid by an altered relationship with protein. In a further paper (Roozemond, 1969*b*) he notes that the quantities of phosphatidyl serine and phosphatidyl ethanolamine present appear to decrease considerably during fixation in formaldehyde or glutaraldehyde. In the case of glutaraldehyde, cross-linking of lipid-NH_2 groups to protein is suggested as a cause. Roozemond (1969*c*) also studied the effects of short-term formaldehyde fixation on phosphatidyl choline; he concluded that fixation for 1.5 hr produced negligible changes in the fatty acid composition, and on the basis of these results, the behaviour of phospholipids during short-term fixation must be attributed, not to the reaction of their unsaturated fatty acids but to reactions of other functional groups present in the molecules. Behaviour of all lipids during longer term fixation may, however, be attributable to reactions of unsaturated compounds with the fixative; the hypothesis has previously been put forward (Jones & Gresham, 1966) that the inability to demonstrate lipid in tissues stored in formaldehyde for a considerable period is due to the conversion of the lipid into a macromolecule which ceases to behave as lipid; macromolecules of a similar type can be produced by the free-radical oxidation of poly-unsaturated fats, either in the body (Jones *et al.*, 1965. 1969) or *in vitro*. These macromolecules are not lipids by definition, because they are insoluble in lipid solvents, but they will stain with the Sudan dyes, giving rise to a paradoxical situation, for Sudan dyes are only soluble in lipids and lipid solvents. It seems probable that if lipids were converted to macromolecules by fixation, these macromolecules also would stain with Sudan dyes, and the absence of staining after prolonged fixation can only suggest that the lipids are no longer present in *any* form in the tissue. They must, therefore, have escaped into the fixative, and regretfully one must realize that the introduction of —OH groups into the lipid molecules probably stimulates the migration of lipids into the fixative before they stand any real chance of being cross-linked and polymerized by reaction with further molecules of $^+CH_2OH$. The presence of phospholipids in the tissues should increase the reaction rate, as even weak emulsifying agents will increase the surface area of an oil droplet many times, and the reactions occurring will be sited mainly at the interface of the oily and aqueous phases. Formaldehyde, as a neutral molecule, would be slightly soluble in fat or oil, but the protonated active form would not be.

The fixative of the future

Both acrolein and glutaraldehyde may lay claim to this title; as general fixatives both are satisfactory, and acrolein could certainly replace formaldehyde as an every-day fixative without real disadvantage (cf. Figs. 24, 25 & 26). As a lipid fixative, acrolein may prove to be more effective than glutaraldehyde.

Under the conditions pertaining to normal fixation, the reactivity of acrolein with respect to alkenic compounds approximates to that of formaldehyde, but preliminary observations on tissues fixed in acrolein, in which Schiff-positive reactions can be obtained in lipid globules, suggest that $^+CH_2$—CH—CHO plays the major role in the reaction. This point is under investigation as the possible basis of a new histochemical reaction for lipids; van Duijn (1961) has proposed an acrolein-Schiff method for protein, and this is further evidence for the predominant role played by $^+CH_2$—CH_2—CHO under 'fixation' conditions.

Numerous authors (Feustel, 1964; Geyer & Feustel, 1966; Feustel & Geyer, 1966; Jones, 1969*b*; Schwab *et al.*, 1970) have previously suggested the use of acrolein as a histological fixative; to quote Feustel & Geyer (1966) 'Als Fixierungsmittel gehört Acrolein in die Gruppe der Lipidstabilisatoren'. Feustel & Geyer consider that acrolein is reactive, has considerable cross-linking capabilities, and forms stable compounds in the tissues. Their studies showed that phospholipids were much better preserved by acrolein than by formaldehyde.

It should be emphasized that the presence of $^{+}CH_2$—CH_2CHO in the fixative renders acrolein as effectively bifunctional as glutaraldehyde. In my opinion, the role of the fixative may have to be extended in the future, and with respect to lipids, the ultimate fixative will not only bind lipid indissolubly to the surrounding protein, but it will also introduce reactive groups into the lipid so that lipids may be identified specifically by histochemical methods. Such a fixative may at present be considered the Philosophers Stone of Histochemistry, but in times when scientists' dreams are seen to become reality all about us, even the histochemist should be permitted an occasional dream.

Acknowledgements

I am indebted to Professors R. I. N. Greaves, Sir Joseph Hutchinson and J. W. L. Beament for the opportunity of working in their Departments.

The loan of equipment by W. G. Pye Ltd is gratefully acknowledged. The studies described in this paper could not have come to pass without the generous use of facilities permitted to me by the University Chemical Laboratory, University of Cambridge, and I thank N. V. Organon for giving me every facility for writing and presenting this paper.

References

ARTUN, T. (1966). Application of the Prins reaction to oleic acid. *J. Am. Oil Chem. Soc.* **43**, 161–4.

BAKER, J. W. (1948). Olefine reactions catalysed by Lewis acids. *Nature (Lond.)* **161**, 171–2.

BEL Le, N. A., LIESEMER, R. N. & MEHMED-BASISCH, E. (1963). The stereochemistry of additions to olefins. III. The Prins reaction with cis- and trans-4-octene. *J. Org. Chem.* **28**, 615–20.

BLOMQUIST, A. T., PASSER, M., SCHOLLENBERGER, L. S. & WOLINSKY, J. (1957*a*). Thermal condensation of formaldehyde with acyclic olefines. *J. Am. Chem. Soc.* **79**, 4972–5.

BLOMQUIST, A. T., VERDOL, J., ADAMI, C. L., WOLINSKY, J. & PHILLIPS, D. D. (1957*b*). Thermal condensation of cyclic olefines with formaldehyde. *J. Am. Chem. Soc.* **79**, 4976–80.

BLOMQUIST, A. T. & WOLINSKY, J. (1957). The mineral acid-catalysed reaction of cyclohexene with formaldehyde. *J. Am. Chem. Soc.* **79**, 6025–30.

BOWYER, D. E., LEAT, W. M. F., HOWARD, A. N. & GRESHAM, G. A. (1963). The determination of the fatty acid composition of serum lipids separated by thin layer chromatography, and a comparison with column chromatography. *Biochem. Biophys. Acta* **70**, 423–31.

DIELS, O. & ALDER, K. (1928). Syntheses in the hydro-aromatic series. I. Addition of 'diene' hydrocarbons. *Annalen Chemie*, **460**, 98–122.

DOLBY, L. J. (1962). The mechanism of the Prins reaction. I. The structure and stereochemistry of a new alcohol from the acid-catalysed reaction of cyclohexene and formaldehyde. *J. Org. Chem.* **27**, 2971–5.

DOLBY, L. J., LEISKE, C. N., ROSENKRANTZ, D. R. & SCHWARTZ, M. J. (1963). The mechanism of the Prins reaction. II. The solvolysis of trans-2-hydroxy-methylcyclohexyl brosylate and trans-2-acetoxy-methyl-cyclohexyl brosylate. *J. Am. Chem. Soc.* **85**, 47–52.

DUIJN van, P. (1961). Acrolein-Schiff, a new staining method for proteins. *J. Histochem.* **9**, 234–41.

ESSENHIGH, D. M., JONES, D. & RACK, J. H. (1964). A sebaceous adenoma; histological and chemical studies. *Br. J. Derm.* **76**, 330–40.

FEUSTEL, E. M. (1964). Zur Eignung der Acroleinfixierung für histochemische Untersuchungen. *Acta Histochem.* **18**, 397–8.

FEUSTEL, E. M. & GEYER, G. (1966). Zur Eignung der Acroleinfixierung für histochemische Untersuchungen. II. Lipide, Enzyme. *Acta Histochem.* **25**, 219–223.

FOLCH, J., LEES, M. & SLOANE-STANLEY, G. H. (1957). A simple method for the isolation and purification of total lipids from animal tissues. *J. Biol. Chem.* **226**, 497–509.

GEYER, G. & FEUSTEL, E. M. (1966). Zur Eignung der Acroleinfixierung für histochemische Untersuchungen. I. Kohlenhydrate, Proteine, Nukleinsäuren. *Z. mikr.-anat. Forsch.* **74**, 392–406.

GIGG, R. & PAYNE, S. (1969). The reaction of glutaraldehyde with tissue lipids. *Chem. Phys. Lipids* **3**, 292–5.

GUSTAVSON, K. H. (1940). The tanning power of aldehydes. *J. int. Soc. Leath. Trades Chem.* **24**, 377–89.

HALL, R. H., STERN, E. S. (1955). Unsaturated aldehydes and related compounds. Pt. VII. Thermal fission of 1,1,3-trialkoxypropanes. *J. Chem. Soc.* 2657–66.

HALLIDAY, N. (1949). The effect of formalin fixation on liver lipids. *J. Biol. Chem.* **129**, 65–9.

HAMMAR, J. A. (1924). Beiträge zur Konstitutionsanatomie VIII. Methode die Menge des Marks, der Rinds und der Rindenzonen, sowie die Menge und Verteilung der Lipoide der menschlichen Nebenniere zahlemässig festzustellen. *Z. Mikr. Anat. Forsch* **1**, 85–190.

HESLINGA, F. J. M. & DEIERKAUF, F. A. (1961). The action of histological fixatives on tissue lipids: Comparison of the action of several fixatives using paper chromatography. *J. Histochem. Cytochem.* **9**, 572–7.

HOPWOOD, D. (1967). Some aspects of fixation with glutaraldehyde. *J. Anat.* **101**, 83–92.

HOPWOOD, D. (1969*a*). A comparison of the cross-linking abilities of glutaraldehyde, formaldehyde, and α-hydroxyadipaldehyde with bovine serum albumin and casein. *Histochemie* **17**, 151–61.

HOPWOOD, D. (1969*b*). The elution patterns of formaldehyde, glutaraldehyde, glyoxal and α-hydroxyadipaldehyde from Sephadex G-10 and their significance for tissue fixation. *Histochemie* **20**, 127–32.

HÜNDGEN, M. (1968). Der Einfluss verschiedener Aldehyde auf die Strukturerhaltung gezüchteter Zellen und auf die Darstellbarkeit von vier Phosphatasen. *Histochemie* **15**, 46–61.

JONES, D., GRESHAM, G. A., HOWARD, A. N. & LLOYD, H. G. (1965). 'Yellow fat' in the wild rabbit. *Nature (Lond.)* **207**, 205–6.

JONES, D. & GRESHAM, G. A. (1966). Reaction of formaldehyde with unsaturated fatty acids during histological fixation. *Nature (Lond.)* **210**, 1386–8.

JONES, D. (1969*a*). The reaction of formaldehyde with unsaturated fatty acids during histological fixation. *Histochem. J.* **1**, 459–91.

JONES, D. (1969*b*). Acrolein as a histological fixative. *J. Microscopy* **90**, 75–7.

JONES, D. & WILLIAMS, D. H. (1969). Characterization of a compound derived from the reaction of formaldehyde with oleic acid, containing hydroxymethyloxy- and hydroxymethyl-groups. *Tetrahedron Letters, No.* 1. 37–8.

JONES, D., GRESHAM, G. A. & HOWARD, A. N. (1969). The aetiology of yellow fat disease in the wild rabbit. *J. Comp. Path.* **79**, 329–34.

KIMMELSTIEL, P. (1929). Über den Einfluss der Formalinfixierung von Organen auf die Extrahierbarkeit der Lipoide. *Z. Physiol. Chem.* **184**, 143–6.

KUTSCHERA-AICHBERGEN, H. (1925). Beitrag zur Morphologie der Lipoide. *Virchows Arch.* **256**, 569–94.

KUTSCHERA-AICHBERGEN, H. (1929). Über der histochemischen Fettnachweis im Gewebe. *Virchows Arch.* **271**, 623–4.

MARE de la, P. B. D. & BOLTON, R. (1966). *Electrophilic Additions to Unsaturated Systems*, p. 198. Amsterdam: Elsevier.

NOZAKI, H., HAMADA, H. & FAGINAMA, I. (1953). Studies on the oil tanning II. *Bull. Nat. Inst. Agric. Sci. (Japan) Ser. G.* **6**, 103–10.

OLSEN, S. (1946). Über die Umsetzung von Formaldehyde mit Cyclohexen und Kenntnis der Hexahydrosalagins. (II. Mitt.) *Z. Naturforsch.* **1,** 671–6.

OLSEN, S. & PADBERG, H. (1946). Über die Umsetzung von Formaldehyde mit Cyclohexen und zur Kenntnis der Prinsschen Reacktion (I Mitt.) *Z. Naturforsch.* **1,** 448–58.

OLSEN, S. (1947). Über die Umsetzung von Formaldehyde mit Ilefinen. *Angew. Chem.* **59,** 32.

OLSEN, S. (1948). Über die mögliche Beteiligung der Formaldehyde-Olefin-reaktion am stofflichen Aufbau der Pflanzen. *Z. Naturforsch.* **36,** 314–20.

PEARSE, A. G. E. (1960). *Histochemistry, Theoretical and Applied,* 2nd Edn., pp. 53–6. London: Churchill.

PRICE, C. C. (1946). *Mechanisms of Reactions at Carbon-carbon Double Bonds.* New York: Interscience, John Wiley.

PRINS, H. J. (1919). Over de condensatie van formaldehyd met onverzadigde verbindingen. *Chem. Weekblad* **16,** 1072–3.

ROOZEMOND, R. C. (1969*a*). The effect of calcium chloride and formaldehyde on the release and composition of phospholipids from cryostat sections of rat hypothalamus. *J. Histochem. Cytochem.* **17,** 273–9.

ROOZEMOND, R. C. (1969*b*). The effect of fixation with formaldehyde and glutaraldehyde on the composition of phospholipids extractable from rat hypothalamus. *J. Histochem. Cytochem.* **17,** 482–6.

ROOZEMOND, R. C. (1969*c*). The effect of calcium chloride and formaldehyde on the fatty acid composition of phosphatidyl choline from cryostat sections of rat hypothalamus. *Histochemie* **20,** 266–70.

SCHWAB, D. W., JANNEY, A. H., SCALA, J. & LEWIN, L. M. (1970). Preservation of fine structures in yeast by fixation in a dimethyl sulphoxide-acrolein-glutaraldehyde solution. *Stain Technol.* **45,** 143–7.

SMISSMAN, E. E. & MODE, R. A. (1957). Stereochemistry of the Prins reaction with cyclohexene. *J. Am. Chem. Soc.* **79,** 3447–8.

SMISSMAN, E. E. & WITIAK, D. T. (1960). Conformational analysis of the Prins reaction. *J. Org. Chem.* **25,** 471–2.

VOGEL, A. I. (1957). *Practical Organic Chemistry,* 3rd enlarged edn., p. 971. London: Longmans.

WHETSTONE, R. R., RAAB, W. J. & BALLARD, S. A. (1951). Condensation products from polyethylenic-unsaturated aldehyde adducts, derivatives thereof, and methods for producing the same. *U.S. Patent* 2,568,426.

WINKLE van, J. L. (1962). Unpublished information, cited in: *Acrolein,* ed. C. W. Smith. New York: John Wiley.

WOLMAN, M. & GRECO, J. (1952). The effects of formaldehyde on tissue lipids and on histochemical reactions for carbonyl groups. *Stain Technol.* **27,** 317–24.

ZIMMERMAN, H. E. & ENGLISH, J. (Jr.) (1953). Stereoisomerism of isopulegol hydrates and some analogous 1,3-diols. *J. Am. Chem. Soc.* **75,** 2367–70.

Theoretical and practical aspects of glutaraldehyde fixation

DAVID HOPWOOD

Department of Pathology,
University of Nottingham

Contents

Synopsis. This review first considers the many structures put forward for glutaraldehyde, and the purification of the commercial material for chemical, histological and histochemical studies. Some practical and theoretical problems of tissue fixation with glutaraldehyde, including artefacts, are then discussed. The chemical reactions with amino acids and proteins are considered next together with the physical changes in the proteins during the reactions. The known reactions of glutaraldehyde with nucleic acids, lipids and mucosubstances are explored briefly.

Introduction

In 1963, Sabatini, Bensch & Barrnett introduced a number of aldehydes as fixatives for studies of cellular ultrastructure. They also investigated the preservation of enzyme activity by these agents. The substances they found most useful were glutaraldehyde, glyoxal, α-hydroxyadipaldehyde, crotonaldehyde, pyruvic aldehyde, acetaldehyde and methacrolein. Since that time glutaraldehyde has found increasing favour as a primary fixative for electron microscopy, usually followed by post-osmication. The other aldehydes they described have found occasional use. Earlier interest in these aldehydes was shown by the tanning industry, and the first synthesis of glutaraldehyde was in 1908 (by Harries & Tank).

The mechanism of the fixative action of these aldehydes, Sabatini *et al.* (1963) suggested, was due to the formation of intermolecular bridges by the condensation of the aldehyde groups with reactive groups in the tissue, analogous to the formation of methylene bridges by formaldehyde. The chemical and physical processes involved in fixation with aldehydes have become better understood since then. The present paper is concerned largely with the mechanisms involved in fixation with glutaraldehyde, both in tissues and in model experiments.

One problem with work on fixation is to define the process. In many ways it is like protein denaturation. Everyone knows what is meant, but cannot put a specific process forward. Negatively, Baker (1960) suggested that tissue fixation prevented autolysis, attack from bacteria and changes in volume and shape. Hayat (1970) suggested that the aims of fixation were rapid preservation of structure with minimum alteration from the living state, and protection during embedding, sectioning and subsequent treatments. Hopwood (1969*a*) also added the rider that there should be no loss of tissue constituents.

One other semantic point needs to be clarified. Is reaction with glutaraldehyde the same as fixation or are they different? I believe that in their extremes they may be different. This difference is probably quantitative rather than qualitative and much valuable information on fixation can be obtained by studying reactions between glutaraldehyde and substances of biological interest, as well as tissues.

The nature of glutaraldehyde

Glutaraldehyde can be bought commercially from many suppliers but apparently is made by only two manufacturers (Hopwood, 1967*b*; Garrett, 1972). The solution is a clear, colourless to pale straw-coloured liquid smelling of rotting apples.

The nature of the solution has been investigated by u.v., i.r., and n.m.r. spectroscopy.

This work suggests that the solution is neither pure nor simple. Richards & Knowles (1967) found that commercial aqueous glutaraldehyde solutions contained polymers largely, together with significant quantities of α,β-unsaturated aldehydes derived by aldol condensation.

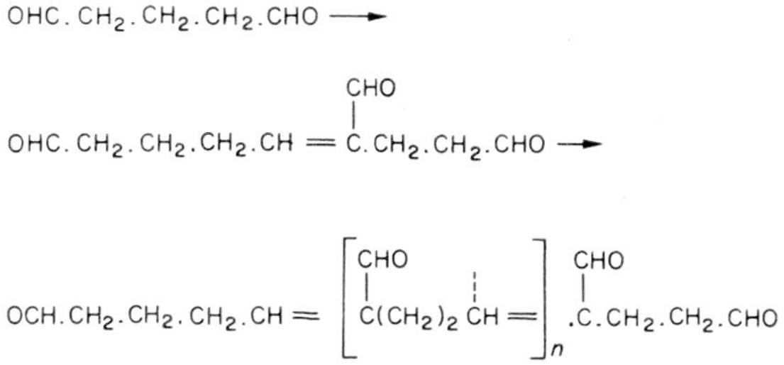

$$\mathrm{OHC.CH_2.CH_2.CH_2.CHO} \longrightarrow$$

$$\mathrm{OHC.CH_2.CH_2.CH_2.CH = \underset{}{\overset{\displaystyle CHO}{\overset{|}{C}}}.CH_2.CH_2.CHO} \longrightarrow$$

$$\mathrm{OCH.CH_2.CH_2.CH_2.CH = \left[\overset{\displaystyle CHO}{\overset{|}{C}}(CH_2)_2\,\overset{\vdots}{CH} = \right]_n \overset{\displaystyle CHO}{\overset{|}{.C}}.CH_2.CH_2.CHO}$$

The condensation products may then undergo cyclization to yield products with six-carbon rings such as:

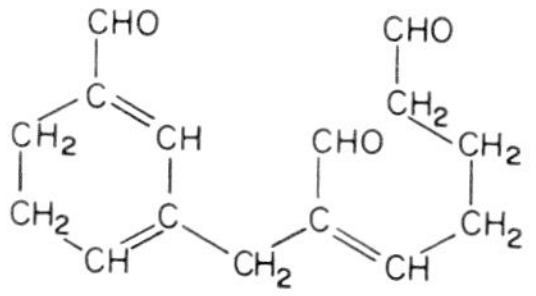

CHO, CHO, C, CH_2, CH_2, CH, CHO, CH_2, CH_2, C, C, CH_2, CH, CH_2, CH

Various other impurities in commercial glutaraldehyde solutions were listed by Anderson (1967) and include acrolein, glutaric acid and glutaraldoxime.

U.v. spectroscopy of glutaraldehyde solutions had previously shown that there were two peaks at 235 and 280 nm (Fahimi & Drochmans, 1965; Hopwood, 1967*b*; Anderson, 1967), the first peak representing the polymer or other impurities and the second peak monomeric glutaraldehyde. Robertson & Schultz (1970), using the Woodward empirical rules, calculated the u.v. absorption maximum for α,β-unsaturated aldehydes derived from glutaraldehyde monomer by the aldol condensation as 231 nm. This value is shifted slightly to the red when water is the solvent. The u.v. absorption coefficient of glutaraldehyde has been found to be temperature sensitive. Hopwood (1967*b*) found that the extinction of material absorbing at 280 nm increased in an exponential manner with temperature. Robertson *et al.* (1970) found the same was true for the 235 nm absorbing material. Various other molecular forms of glutaraldehyde have been put forward by Japanese workers. Aso & Aito (1962) found that in aqueous solutions glutaraldehyde existed as tetramers and pentamers and that cyclopolymerization is possible.

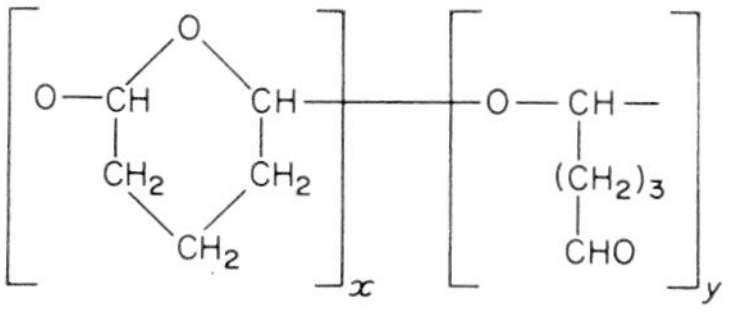

O, O—CH, CH, CH_2, CH_2, CH_2; O—CH—, $(CH_2)_3$, CHO

$$[\ldots]_x\,[\ldots]_y \qquad x : y = 3 : 8$$

Yokota *et al.* (1965) found that cyclic forms were temperature dependent. Below 0°C, form (I) predominates and above 0°, form (II) predominates.

$$\text{I} \qquad -\left[-OCH_2(CH_2)_3CO-\right]-\left[-OCH_2(CH_2)_3O-\right]-\left[-CO(CH_2)_3CO-\right]- \quad \text{II}$$

Evidence has also been brought forward by Hardy *et al.* (1969), also using n.m.r. spectroscopy, that in aqueous solutions glutaraldehyde exists as a mixture of hydrated forms of the monomer in equilibrium with unhydrated forms as follows.

$$OCH.CH_2.CH_2.CH_2.CHO \underset{-D_2O}{\overset{+D_2O}{\rightleftharpoons}} \text{cyclic } (e)H_2C-CH_2(e)-CH_2(e),\ (b)HC(OD)-O-CH(OD)(b)$$

$$\text{(DO)}_2\text{HC(c)}-\text{(e)}H_2C-CH_2(e)-CH_2(e)-CH(OD)_2\text{(c)} \underset{+D_2O}{\overset{-D_2O}{\rightleftharpoons}} CHO\text{(a)}-\text{(d)}H_2C-CH_2(e)-CH_2(e)-CH(OD)_2\text{(c)}$$

These workers determined the equilibrium constants between the various molecular forms and found that very little free glutaraldehyde exists in solution. The cyclic monohydrate was easily formed, but if a solution of distilled glutaraldehyde was stored for several weeks near neutrality, little polymerization occurred.

Purification

Various techniques have been proposed for the purification of glutaraldehyde. This was prompted by the variability of fixation, especially for enzyme electron histochemistry, after using commercial glutaraldehyde. Fahimi *et al.* (1965) put forward the first technique; they developed a system of vacuum distillation which yielded a product of high concentration and purity as assessed by u.v. spectroscopy. Smith & Farquhar (1966) suggested distillation at atmospheric pressure as a method of purification. A simpler technique was introduced by Anderson (1967). He found that if commercial glutaraldehyde was treated with charcoal then the 'impurities' could be removed and reliable fixation ensues. The impurities were probably polymers as the u.v. spectrum of the charcoal-treated glutaraldehyde lacked the peak at 235 nm. Hopwood (1967*b*) found that substances giving rise to the two u.v. absorbing peaks could be separated by subjecting various commercial glutaraldehydes to chromatography on Sephadex G-10. The aldehyde activity closely paralleled the elution of the material with an extinction at 280 nm. This fraction had a smaller elution volume than the 235 nm absorbing material. Gel filtration normally separates substances according to molecular weights, the higher molecular weights appearing at the lower elution volumes. However, cyclic substances have been shown to be retained on Sephadex columns by Gelotte (1960). Hopwood

(1967*b*) also reported that the elution volume of glutaraldehyde from Sephadex was pH sensitive, being increased as the pH is lowered. This could be due to a molecular change which depends upon the hydrogen ion concentration. Purification will often lead to a decrease in the concentration of aldehyde present in the solution.

Evans (1969) has used a batch extraction method for the purification of commercial glutaraldehyde. He added anhydrous calcium chloride and afterwards extracted the mixture with diethyl ether. The ether extracts were dried, the ether removed and the remaining liquid distilled and the fraction boiling at 82.5—84°C at 15 mm Hg recovered.

Inorganic ions are present in some batches of glutaraldehyde. These include ferric and cupric ions at several parts per million.

Carstensen and his colleagues (1971) have commented that according to ultracentrifugation studies the material in glutaraldehyde with an absorbance maximum at 235 nm is either a dimer or a higher polymer(s).

One thing probably observed by most investigators who have distilled glutaraldehyde is that a glassy solid remains in the distillation flask. Hardy *et al.* (1969) thought that this was polymerized glutaraldehyde, probably in the cyclic form suggested by Yokota *et al.* (1965). If this form was heated it reverted to a monomeric form.

Estimation of glutaraldehyde

A number of methods have been used for the assay of glutaraldehyde. Fahimi *et al.* (1965) suggested the use of absorption spectrophotometry at 280 mm for the distilled commercial aldehyde. This obviously has certain limitations as a method. Fein & Harris (1962) used an iodometric method based on the following reactions.

$$R.CHO + NaHSO_3 \longrightarrow RCHO.NaHSO_3$$
$$RCHO.NaHSO_3 + 2I_2 + 3Na_2CO_3 \longrightarrow R.COONa + Na_2SO_4 + 4NaI + H_2O + 3CO_2$$

A shorter variation of this method, which has been used to assay various biologically important aldehydes and ketones (Clift & Cook, 1932), has been used by Hopwood (1967*c*). The N-methylbenzothiozone method of Blumenfeld *et al.* (1963) has also been used (Hopwood, 1967*b*). Anderson (1967) analysed glutaraldehyde by the hydroxylamine method as do Union Carbide, but Frigerio & Shaw (1969) thought the method too cumbersome and time-consuming. These last workers returned to an iodometric method which they found gave satisfactory results. Hardy *et al.* (1969) removed impurities from commercial glutaraldehyde solutions by first treating the solution with sodium chloride and ethanol followed by vacuum distillation of the supernatant. They went on to criticize the method of Fahimi *et al.* (1965) on the grounds that the u.v. extinction coefficients were functions of time, temperature and initial concentration so that Beer's law was not obeyed.

Storage of glutaraldehyde

If commercial glutaraldehyde is stored for a long time a precipitate often forms. This and the variable results of tissue fixation has lead to an investigation of the effects of storage on commercial glutaraldehyde solutions. Frigerio *et al.* (1969) found that if 50% glutaraldehyde solution was stored at 1°C for 341 days the aldehyde titre decreased only by 2.46%. Light and air had relatively little effect at room temperature over four days. Ferrous, ferric and cupric ions (10^{-4} M) had no effect either. They concluded that the formation of a polymer, with a low aldehydic activity was inevitable. The optimum concentration for stability was 2–10%. The formation of polymer appears to be proportional to hydrogen ion concentration (Hardy *et al.*, 1969). Under alkaline conditions at pH 8, glutaraldehyde polymerizes rapidly to give a white precipitate. Polymerization may also occur under acid conditions. This may revert to a monomer on heating. Trelstad (1969) also investigated the formation of polymer in distilled glutaraldehyde which had been stored at various hydrogen ion concentrations for more than a year. Judged by changes in the ratio of the extinctions at 250 and 280 nm, glutaraldehyde stored at pH 5 kept the best. Anderson (1967) also recommended storage at 4°C but under nitrogen or Freon. Some of the commercial practices, e.g. storage in the presence of barium carbonate, would seem to have no sound basis.

Tissue fixation with glutaraldehyde

The practical aspects of tissue fixation with glutaraldehyde are dealt with in various practical books on electron microscopy (e.g. Hayat, 1970; Kay, 1967). There are some aspects, however, relevant to the themes of this paper. A brief discussion of them follows.

(i) *Penetration*

Glutaraldehyde penetrates into tissues relatively slowly (Hopwood, 1967*a*; Ericsson & Biberfeld, 1967; Chambers *et al.*, 1968) although the rate depends on temperature.

Bishop *et al.* (1968) cross-linked large crystals of β-lactoglobulin. They found that if the reaction time was limited, then the centre of the crystal remained soluble and they were left with a cross-linked shell. The slow rate of penetration may be associated with diffusion artefacts. Glutaraldehyde reacts quickly, however, as Flitney (1966) has shown.

Because of the slow rate of penetration of glutaraldehyde into tissues, many workers use vascular perfusion rather than immersion as the technique of choice. Various descriptions of the technique are to be found, e.g. Forssmann *et al.* (1967). Perfusion can give rise to various artefacts; for example in the liver the rise of too high a pressure can cause the formation of clear vesicles in the hepatocytes (Hopwood, 1972*b*). Gil & Beibel (1971) have described an apparatus which gives a constant controllable hydrostatic pressure.

(ii) *Comparison of fixatives*

There have been a number of papers in which the ultrastructure of various tissues fixed by glutaraldehyde have been compared with that obtained with other fixatives, notably osmium tetroxide. The earlier papers have been reviewed by Hopwood (1969) but outstanding amongst these are the reports by Trump & Ericsson (1965) and Wood & Luft (1965). More recently, Landon (1970) has investigated the fine structure of rat striated muscle and concluded that it is dependant on the manner of fixation. Osmium tetroxide-fixed material showed a square or woven lattice with a spacing of 22 nm and its axes offset by 45° to the axes of the I filaments. Glutaraldehyde-fixed muscle had a square lattice pattern with a spacing of 11 nm. This was probably due to superimposition of two square lattices of 22 nm period. The importance of this variation in structure with fixation need not be stressed further. Beauvillain (1970) compared the effects of glutaraldehyde, osmium tetroxide and a glutaraldehyde-formaldehyde mixture on the fine structure of the mouse median eminence. The mixture was found to give the most reproduceable results. Hardin & Spicer (1970) have compared the fine structure of rat trigeminal ganglia fixed with glutaraldehyde and pyroantimonate-osmium tetroxide. The latter showed a precipitate in the nucleoli which may represent sodium ions.

(iii) *Effects of osmolarity of the fixative solution*

Commonly-used fixative solutions of glutaraldehyde also contain a phosphate or cacodylate buffer that increases the osmolarity of the solution. The effects of fixative osmolarity on the ultrastructure produced have been studied by a number of workers. Isotonic and hypotonic solutions were found to give poor results by perfusion and immersion fixation. Slightly hypertonic solutions were favoured although in the central nervous system they could produce large extracellular spaces (Fahimi & Drochmans, 1965*b*; Schultz & Karlsson, 1965; see also Hopwood, 1969*a*).

Webster & Ames (1969) and with their colleague Nesbett (1969) commented on the shrinkage in cellular components produced by 3% glutaraldehyde in 0.05 M phosphate buffer compared with rabbit retina fixed in isotonic osmium tetroxide solution. They also investigated the effects of sodium chloride. 'Hypertonic' fixation produced decreases in the linear dimensions, areas and volumes of retina cells. The mitochondria behaved like osmometers, although nuclei did not.

Bohman & Maunsbach (1970) investigated the effects on tissue ultrastructure by altering the colloid osmotic pressure in rat kidney, pancreas and intestine fixed by perfusion with glutaraldehyde. The only variable was the concentration of dextrans (mol. wt. 40000 or 80000) and polyvinylpyrrolidine (mol. wt. 40000). Tissues fixed in the absence of polymers had enlarged extracellular spaces compared with those fixed in their presence (Fig. 1). They concluded that the osmotic pressure of fixatives should be carefully controlled when structures or cellular relationships in labile extravascular spaces were to be investigated. Previously, dextran has been found useful in the preservation of myocardium (Rostgaard & Behnke, 1965) and detergent-treated chloroplasts (Deamer & Crofts, 1967).

(iv) *Glutaraldehyde-containing fixative mixtures*

Although glutaraldehyde fixation has many points in its favour, certain disadvantages have been noted. Curgy (1968), for example, found myelinization of lipids and Franke *et al.* (1969) reported alterations in microtubules and a dispersed pattern of ribosomes.

Attempts have been made to overcome these difficulties by using a mixture of fixatives either empirically or by design. Thus, in a study on seeds, which are very difficult to fix for electron microscopy, Mollenhauer & Totten (1971) recommended as a primary fixative a mixture of glutaraldehyde, formaldehyde generated from paraformaldehyde, and acrolein. Respectively, these provided maintenance of shape, stabilization of the protein and rapid penetration. This mixture was followed either by post-osmication or treatment with potassium permanganate.

A simpler, much used mixture is that of glutaraldehyde-formaldehyde or formaldehyde made from paraformaldehyde which Karnowsky (1967) introduced. Bloem (1968) hydrolysed leather tanned with a mixture of formaldehyde and glutaraldehyde. At alka-

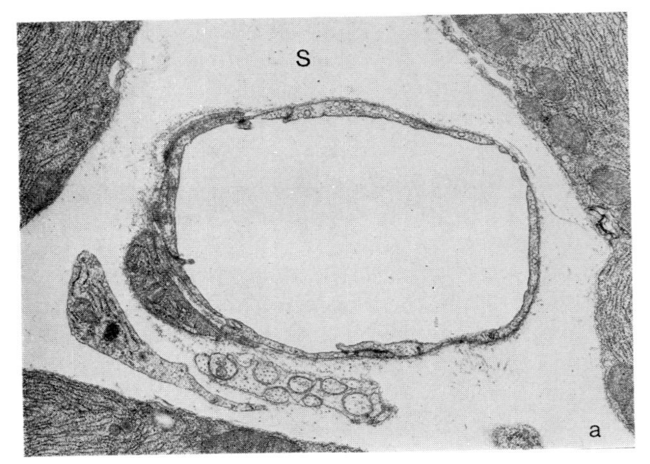

Figure 1a. Rat pancreas perfused with glutaraldehyde. An almost empty space (S) is present between vessel and epithelial cells. × 1900

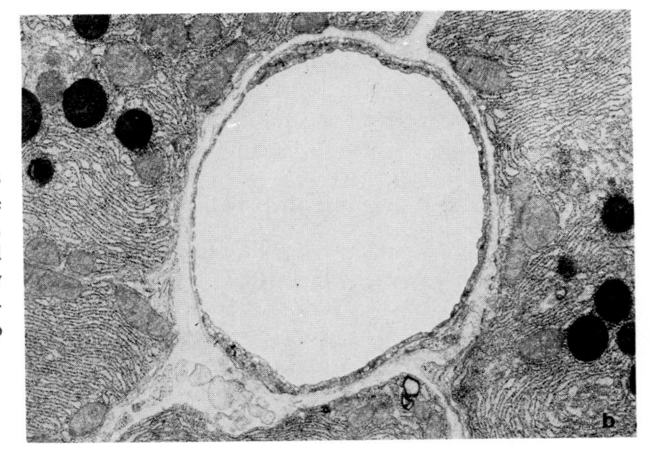

1b. Blood vessel in pancreas perfused with glutaraldehyde fixative containing dextran. The space between the vessel and epithelial cells is narrow and contains some extracellular material. × 17000

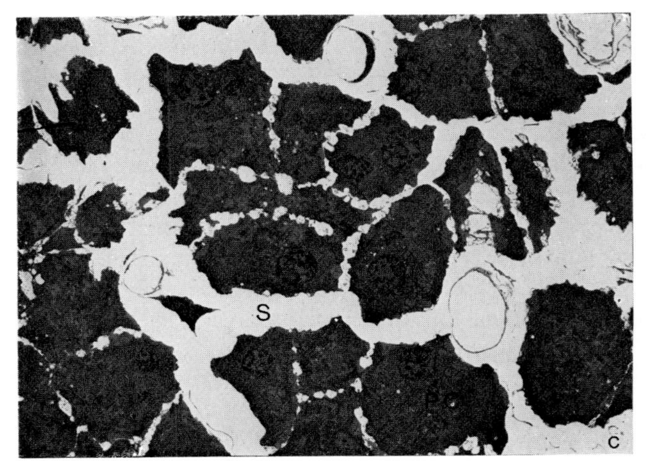

1c. Pancreas perfused with glutaraldehyde made hypertonic with sodium chloride. Note wide extracellular spaces (S) and dense cytoplasm of exocrine cells (PC) and separation of cells within acini. From Bohman *et al.* (1970) with the authors' permission. × 2100

line pH levels he suggested that the two aldehydes may react together to form 1,3-cyclohexandione.

$$H_2C(CH_2{-}CHO)_2 + O{=}CH_2 \rightarrow H_2C\langle CH_2{-}CO{-}CH_2{-}CO\rangle CH_2 + H_2O$$

Other workers have recommended various glutaraldehyde-osmium tetroxide mixtures. Some of the earlier attempts were unsuccessful (Daneel & Weissenfels, 1965). Later reports have been of good or improved fixation compared with the usual glutaraldehyde-postosmication routine (Trump & Bulger, 1966; Hirsch & Fedorko, 1968; Bodian, 1970; Hecker, 1970). Mixtures of osmium tetroxide and glutaraldehyde turn black and Baker (1960) warned about the non-compatability of osmium tetroxide and aldehydes.

This problem was investigated by Hopwood (1970) who studied the reactions between osmium tetroxide and formaldehyde and glutaraldehyde in tissues and by various physical techniques. In solution the aldehyde is oxidized to the appropriate acid and the osmium tetroxide reduced to osmium black, an ill-defined group of polymeric substances (Hanker *et al.*, 1967). In agreement with Bahr (1954), Hanker *et al.* (1967) found that the rate of formation depended upon the temperature and concentrations of glutaraldehyde and protein; low concentrations of reagents reduced the amount of pigment formed (Franke *et al.*, 1969). The reaction between glutaraldehyde and osmium tetroxide was analysed by spectroscopy. The form of the reaction appeared to be one in which the two reactants formed an intermediate compound which then broke down to form osmium black (Figs. 2, 3). This form is closely analogous to the theoretical one (Cleland, 1967) for this scheme of reactions.

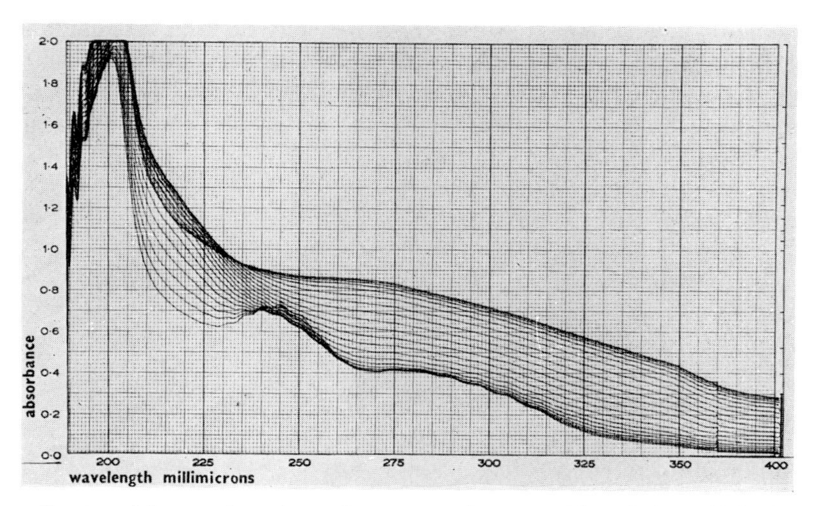

Figure 2. Spectra of the reaction mixture between osmium tetroxide and glutaraldehyde at 25°C. Time between spectra, 10 min. Note increase in absorbance (extinction) at 400 nm and rise and subsequent fall in 205–230 nm region. From Hopwood (1970).

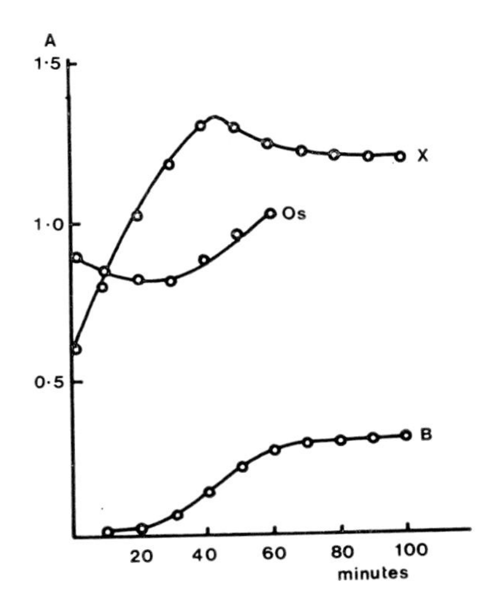

Figure 3. Change in extinction (A) with time of a glutaraldehyde-osmium tetroxide reaction mixture. B = osmium black measured at 400 nm; Os = osmium tetroxide measured at 245 nm; X = intermediate product measured at 215 nm. No corrections applied. From Hopwood (1970).

The result of the reaction between osmium tetroxide and glutaraldehyde has been demonstrated by viscosimetry (Hopwood, 1967). The viscosity changes produced between bovine serum albumin and osmium tetroxide formaldehyde or glutaraldehyde were measured. Mixtures of fixatives were also investigated. The final viscosity was found to be less than that of the more active of the pair, probably because of the competition for similar sites on the protein molecules. Polyacrylamide gel electrophoresis experiments indicated that glutaraldehyde was the best at cross-linking of bovine serum albumin after 1 hr.

These experiments were then repeated with blocks of tissue that were fixed for 24 hr with osmium tetroxide or with formaldehyde or glutaraldehyde. The depth of penetration of the osmium tetroxide was then measured (Fig. 4) and was in the order osmium tetroxide alone, osmium tetroxide-formaldehyde, and osmium tetroxide-glutaraldehyde. In the last mixture the centre of the block was fixed by glutaraldehyde only. If the amount of osmium tetroxide is not infinite but the pieces of tissue large, then after a finite time the depth penetrated is related to its concentration.

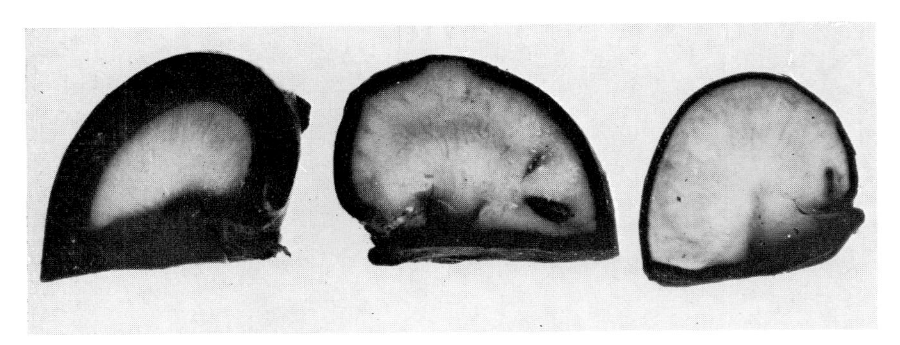

Figure 4. Transected rat kidneys fixed for 24 hr at 4°C. Magnification × 3.3 (*a*) osmium tetroxide; (*b*) osmium tetroxide plus formaldehyde; and (*c*) osmium tetroxide plus glutaraldehyde. From Hopwood (1970).

The use of mixtures of fixatives for electron microscopy has become increasingly popular. In the model situation where the fixatives and protein can mix freely, the cross-linking is less than the most efficient of the pair. However, in tissue fixation there is also the problem of penetration of the components. In the case of a glutaraldehyde-formaldehyde mixture, the former has a low diffusion constant value and the latter a high one. Thus the formaldehyde will stabilize structures until the glutaraldehyde diffuses in. The early reactions of formaldehyde with tissues are apparently reversible (Pearse, 1968) whereas those of glutaraldehyde are not (Hopwood *et al.*, 1970).

Schwab *et al.* (1970) have introduced a dimethyl sulphoxide-acrolein-glutaraldehyde mixture which they recommend for yeast fixation. They suggested that the dimethyl sulphoxide acted as a permeating agent allowing the maximum penetration of the cell wall.

Recently, glutaraldehyde mixtures have been introduced for the staining of the glycocalyx. Shea (1971) recommended a mixture of glutaraldehyde with either Alcian Blue 8GX or cetylpyridinium chloride which he found improved the staining with lanthanum. Behnke & Zelander (1970) came to similar conclusions. Some reaction may take place with these nitrogen-containing compounds and glutaraldehyde.

(v) *Artefacts*

Glutaraldehyde may cause various artefacts as shown, for example, by the work of Peters & Ashley (1967). Using autoradiography they compared the effects of glutaraldehyde, osmium tetroxide and formaldehyde in chemically binding [^{3}H]leucine to bovine serum albumin and to tissues such as liver in the presence of puromycin to inhibit protein synthesis. Glutaraldehyde bound 25% of the [^{3}H]leucine to the bovine serum albumin but formaldehyde bound only 0.5%. In the liver slices fixed with glutaraldehyde, 63% of the grains were due to binding of the free amino acids. The percentage of grains in the case of osmium tetroxide and formaldehyde was 25 and 4 respectively. Peters & Ashley, therefore, recommended formaldehyde as a fixative for autoradiography and added that the binding of radioactive substances by the fixative under the specific experimental conditions should be determined. Similarly Hodson & Marshall (1967) have reported that [^{3}H]tyrosine is bound to rabbit retina by glutaraldehyde in the presence of puromycin and also to rabbit serum albumin.

Equally, artefacts may be produced by the loss of labelled and non-labelled macromolecules during fixation and preparation for microscopy. Vanha-Perttula & Grimley (1970) have investigated this problem quantitatively using human carcinoma monolayers and [^{3}H]-labelled substances. Under conventional conditions the losses during fixation and washing amounted to 15%. Losses of proteins and amino acids were greater with formaldehyde than with glutaraldehyde fixation. Both fixatives also produced some non-specific binding of [^{3}H]amino acids. There was less non-specific binding of [^{3}H]mannose, [^{3}H]thymidine and [^{3}H]uridine. The relationship of time of fixation, buffers, fixative concentration and temperature were also considered. These findings are obviously significant in the quantitative interpretation of high resolution autoradiography.

At variance with these findings is the report by Descarries & Droz (1970) who found, using [^{3}H]noradrenaline, that glutaraldehyde was the best fixative for autoradiography of brain. They found, surprisingly, that the [^{3}H]noradrenaline was not associated particularly with the large granular vesicles, although such an association has been reported. Perhaps their random distribution may in part be artefactitious.

Another artefact associated with glutaraldehyde fixations has been investigated by Reale and his colleagues (Reale & Luciano, 1970; Thiessen *et al.*, 1970). This problem is associated with the relatively slow penetration of glutaraldehyde into tissues. The fixation in these experiments lasted 2 hr in glutaraldehyde. These workers found that [^{59}Fe]-labelled haemoglobin diffused from the unfixed centre of blocks of spleen ($2 \times 2 \times 5$ mm) to the periphery. During subsequent post-osmification there was a further loss of [^{59}Fe]haemoglobin from the centre of the block. This underlines the need to use small blocks of tissue for fixation.

Shrinkage of tissues is a common artefact occurring during fixation. This has been investigated in tissues and other experimental models (Hopwood, 1967*a*). The reduction in volume of tissues was 6% after 18 hr fixation at 4°C with 4% glutaraldehyde.

Jahn (1971) has warned against the use of glutaraldehyde for any histochemical technique where aldehydes are generated, e.g. the periodic acid-Schiff and Feulgen hydrolysis techniques as false positives will occur. Hopwood (1967*a*) had demonstrated this previously.

(vi) *Agonal changes*

When tissues are removed from a living animal the agonal changes may be divided into two phases:

(1) from the removal of the tissue from its blood supply to putting it in the fixative solution, and

(2) from the onset of fixation until the fixative has penetrated into all the cells to a concentration at which they become fixed.

The first phase is easily controlled although there may be difficulties with surgical specimens. During the second phase glutaraldehyde will diffuse into the tissue forming a concentration gradient. Information on the effects of low concentrations of glutaraldehyde on organelles will be of value in interpreting cytological findings. Some of the problems and consequences of these phenomena have been discussed recently (Hopwood, 1972*a*).

(vii) *Cell and other membrane systems*

Many fixatives have considerable effects on cell membranes, both on their morphological and physiological properties. Altered function is probably best understood in relation to mercurial fixatives where changes in enzyme activity have been extensively investigated (for review, see Hopwood, 1972*a*).

The effects of glutaraldehyde have been investigated only to a small extent. In 1966 Jard *et al.* investigated the action of a number of commonly-used fixatives on the ultrastructure and sodium and water transport of frog urinary bladder. They found some fixatives (e.g. ethanol, potassium permanganate) rendered epithelium completely permeable. Other fixatives, e.g. osmium tetroxide, made the epithelium permeable to water but not to sodium ions. A third group of fixatives preserved part of the impermeability to both. Examples of this group are glutaraldehyde and formaldehyde. Jard *et al.* (1966), however, found that ultrastructural preservation and functional integrity did not concur. These permeability changes in toad bladder were confirmed by Grantham *et al.* (1971).

They also found the bladder remained permeable to water 24 hr after treatment with glutaraldehyde. Carstensen *et al.* (1971) found that after red blood cells had been fixed by glutaraldehyde they ceased to act as osmometers (but see below and section on *mitochondria*).

Van Harreveld & Khattab (1968) investigated the conductivity of the cerebral cortex during glutaraldehyde fixation. They had become interested in water distribution in the central nervous system and the effects of asphyxia and found that during fixation extracellular material moved into cells giving an enlarged, electron-lucent appearance. This has also been noted by Majno and his colleagues in their studies on cerebral anoxia (Chiang *et al.*, 1968). During perfusion of the brain with glutaraldehyde the conductivity of the cerebral cortex drops by 60%. Perfusion of asphyxiated tissues with glutaraldehyde does not produce a further drop. Osmium tetroxide causes a marked rise in conductivity of both anoxic cortex and glutaraldehyde-perfused cortex. These workers interpreted these findings as indicating that glutaraldehyde caused a transport of extracellular material, including chloride ions, into the cells, similar to asphyxia. Van Harreveld & Steiner (1970) confirmed these changes in extracellular space using freezing and ethanol substitution on brain whose circulation had been stopped for 30 sec or after 8 min anoxia.

Heller *et al.* (1971) investigated the behaviour of the discs in the outer segments of retinal rods in the presence of various electrolytes. The discs acted as osmometers showing swelling and contraction according to the Boyle-van't Hoff law. The discs were permeable to acetate and ammonium ions but were impermeable to sucrose and the following ions: sodium, potassium, magnesium, calcium, chloride and phosphate. When the discs had been fixed in glutaraldehyde they were still able to undergo volume change.

Bone & Denton (1971) have investigated the effects of various fixatives on the reflecting scales of the herring. In agreement with Jard *et al.* (1966) they found that fixation with aldehydes did not destroy the cellular osmotic behaviour. If the aldehydes were made up in sucrose or salt solution of about 60% of that to which the cells are exposed in life there is no volume change. These changes are discussed in greater detail by Bone & Ryan (1972).

Ellar *et al.* (1971) have investigated the effects of 0.2–0.5% glutaraldehyde on a number of plasma membrane-bound enzymes in *Micrococcus lysodeikticus*. They found that washing membranes would remove Ca^{2+}-dependant ATPase and NADH dehydrogenase. This could be prevented if the membranes were first treated with glutaraldehyde. Glutaraldehyde-treated membranes also contain more protein than untreated ones. Further support that glutaraldehyde binds the enzymes to the membranes was deduced from electron microscopy. These workers used negatively stained membranes and found that successive washes of the untreated membranes progressively reduced the granular substructure on the membrane surface. Membranes treated with glutaraldehyde retained the granular fine structure despite the washing procedures (Figs. 5, 6, & 7).

(viii) *Mitochondria*

A certain amount of information is now available on the effects of low and high concentrations of glutaraldehyde on isolated mitochondria. Packer *et al.* (1968) used high concentrations of glutaraldehyde to achieve gross ultrastructural fixation in mitochondria undergoing rapid volume changes associated with energized ion transport. They cor-

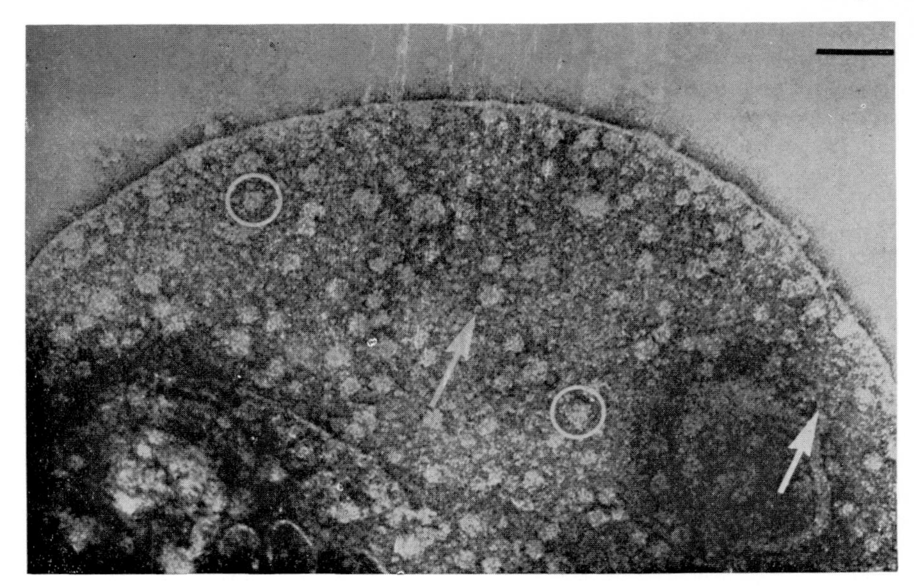

Figure 5. Untreated *M. lysodeikticus* membranes before washing. Circle contains aggregates of smaller units (ATPase). Negatively stained with ammonium molybdate. Marker corresponds to 0.1 μm. From Ellar *et al.* (1971) with the authors' permission.

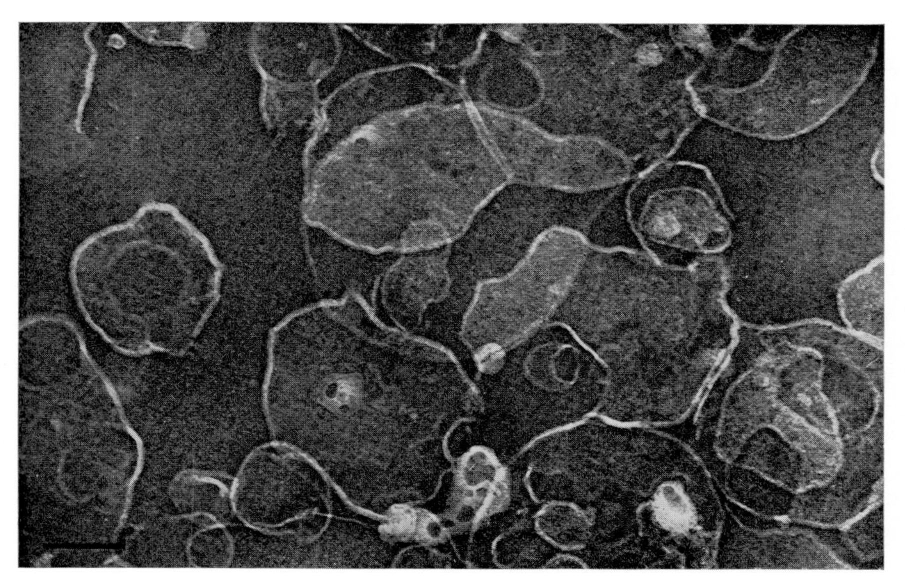

Figure 6. Membranes of *M. lysodeikticus* after six consecutive washes. Marker corresponds to 1 μm. From Ellar *et al.* (1971) with the authors' permission.

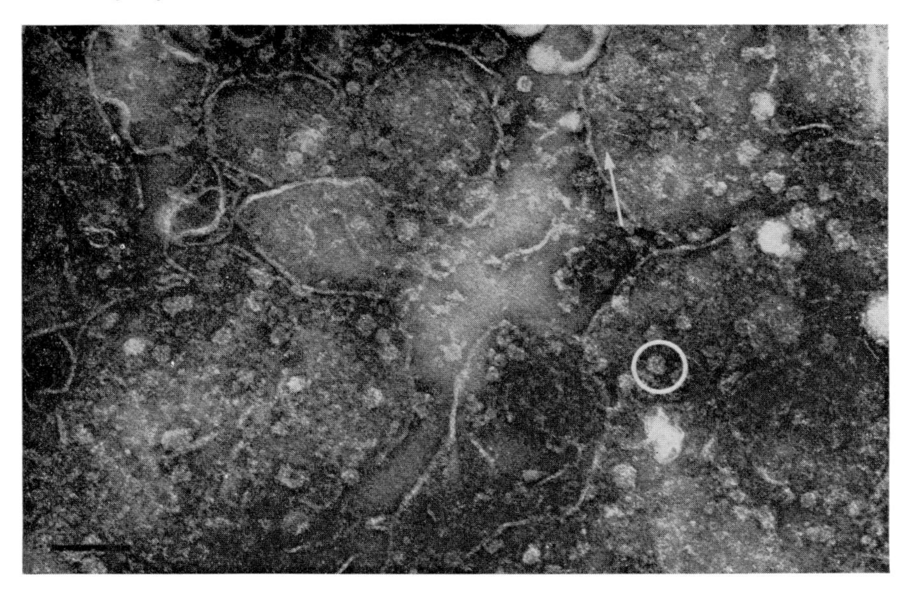

Figure 7. Membranes of *M. lysodeikticus* treated with 0.5% glutaraldehyde and then washed six times. Marker corresponds to 1 μm. From Ellar *et al.* (1971) with the authors' permission.

related their measurements with electron microscopy. Volume changes associated with ion movements reflected changes in the inner membrane compartment. Packer & Greville (1969) found that it was possible to fix the mitochondria and yet preserve the electron transport and other functional systems. They used as a criterion of fixation the

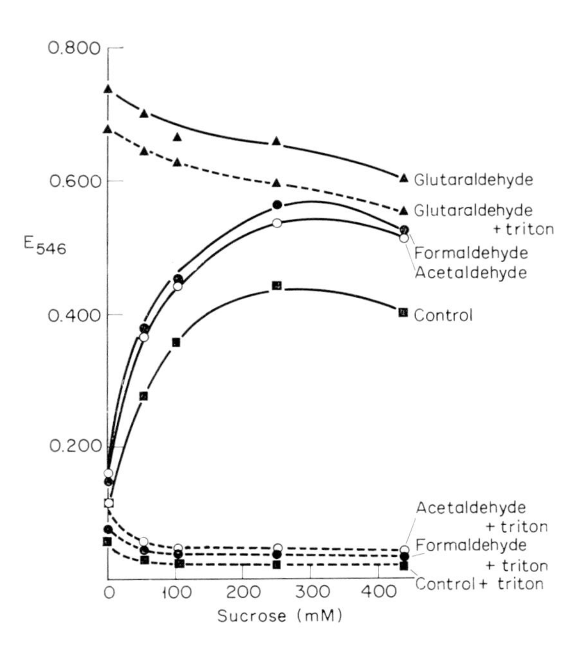

Figure 8. Test for fixation of mitochondria by aldehydes. The mitochondria were suspended (20 mg protein/ml) in a medium containing 0.44 M sucrose, 1 mM EDTA and 50 mM aldehyde and left for 15 min at room temperature. 0.1 ml aliquots were then diluted 40-fold with sucrose solutions to give the final molarities shown. 10 min later the extinction (F_{546}) at 546 nm was measured. Then 0.02% Triton X-100 was added to each sample and after a further 10 min the extinction was measured again. From Packer *et al.* (1969) with the authors' permission.

prevention of change in volume of mitochondria brought about by Triton X-100 (Fig. 8). The minimum concentration of glutaraldehyde needed to achieve this was 5 mM for a reaction with a protein at a concentration of 2 mg/ml. Malondialdehyde did not fix mitochondrial configurations (Horton & Packer, 1970*b*).

In a later study Wrigglesworth *et al.* (1970) suggested that glutaraldehyde fixation of mitochondria preserved the gross ultrastructure but prevented the resolution of fine details. However, they found fine resolution was possible by freeze etching. These workers also investigated the ability of mitochondria to metabolize glutaraldehyde. They found that the metabolism occurred rapidly at low concentrations (2 mM), the glutaraldehyde acting as a substrate along with oxygen (Fig. 9). Horton & Packer (1969) investigated the ability of mitochondria to react also with formaldehyde, malonaldehyde, acetaldehyde, glyoxal and butyraldehyde. They concluded there was probably an aldehyde oxidase in mitochondria of low specificity. Mg^{2+} and Mn^{2+} were stimulatory but inorganic phosphate and ATP were neither essential nor stimulatory.

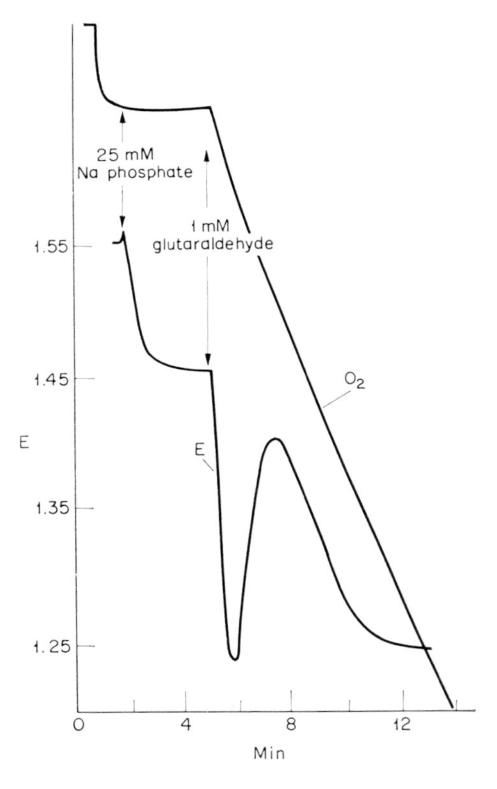

Figure 9. Glutaraldehyde oxidation and mitochondrial swelling. The reaction mixture contained initially 121 mM sucrose and 0.5 mM EDTA. Other additions shown by arrows; temperature, 22°C; pH 7.8. From Packer *et al.* (1969) with the authors' permission.

Elsewhere Horton & Packer (1970) pointed out that under natural conditions when lipid peroxidation occurs, malonaldehyde is one of the products. This in turn can react on mitochondrial membranes. This may be a step in the formation of the ageing pigment lipofuscin. Milch (1963) suggested that glutaraldehyde may also play a part in tissue aging. The dialdehyde may be produced by intermediary metabolism and react with collagen fibres.

Fortes (1971) showed that when glutaraldehyde was added to mitochondrial suspensions, hydrogen ions are generated proportional to the mitochondrial protein (about 250 μmol/g). He also showed that the rate of hydrogen ion formation depended on the metabolic state of the mitochondria. This may affect the conformation of the mitochondrial proteins and the availability of amino groups for reaction. The state of ionization of the proteins is also important in this respect. In detergent-solubilized mitochondria, hydrogen ion production increases exponentially with the pH of the medium. Under the anaerobic conditions which hold during fixation, tissues are known to become acidic.

(ix) *Chloroplasts*
Packer (1968*a*) investigated the effects of light induced-transport in glutaraldehyde-fixed spinach chloroplasts, which retained the capacity to manifest light-dependent hydrogen ion uptake. The movement of other ions such as weak acid anions or weak base cations were largely inhibited because fixed chloroplasts lost their capacity to undergo changes in volume by an osmotic mechanism. Increased stability of glutaraldehyde-treated chloroplast against degradation by a number of enzymes has been noted previously by Mantai (1970) and West *et al.* (1970).

(x) *Reaction with micro-organisms*
Glutaraldehyde is active against viruses, fungi, bacterial spores and vegetative cells, which has lead to its marketing as a sterilizing agent for surgical instruments ('Cidex'—a 2% solution of glutaraldehyde). Rubbo *et al.* (1967) have reviewed its biocidal activities. It is more active under alkaline conditions than acid and this is associated with a red colouration of some bacteria (Munton & Russell, 1970, 1971). This red colouration was associated with the non-mucopeptide layers cell wall and not the cytoplasmic constituents. After reaction of the bacteria with glutaraldehyde, the cell walls were mechanically strengthened.

Reaction of glutaraldehyde with substances of biological interest

(a) AMINO ACIDS AND OTHER LOW MOL. WT. SUBSTANCES
There has been no systematic study of the reaction of glutaraldehyde with amino acids as has been reported earlier for formaldehyde (French & Edsall, 1945). However, there is data available for some amino acids: lysine (Quiocho & Richards, 1966; Bowes & Cater, 1968; Alexa *et al.*, (1971) tyrosine, tryptophan and phenylalanine (Hopwood *et al.*, 1970; Bowes *et al.*, 1968); histidine, proline, serine, glycine, glycylglycine and arginine (Chisalita *et al.*, 1971; Alexa *et al.*, 1971). The authors cited investigated the ability of a number of aldehydes to react with the animo acids at various pH levels from 2 to 11. They ranked the reactive end-groups of the amino acids in decreasing order of reactivity: ε-NH_2, peptide, guanidine, secondary amino and hydroxyl. Further, they found that the reaction constants for glutaraldehyde and glycine, serine and proline had maxima in the range pH 6–7. The reactions constants with lysine and arginine simply increased with pH. For glycylglycine there were two maxima at about pH 5.5 and 9.5.

Hughes & Thurman (1970) investigated the reaction of glutaraldehyde with the cell walls of *Bacillus subtilis*. They found that a reaction occurred with the ε-amino groups of diaminopimelic acid. Evans (1969) has shown that glutaraldehyde reacts with various sulphydryl- and hydroxyl-containing compounds and with compounds with active hydrogens on nitrogen and carbon atoms.

The following biogenic amines have been found to react with glutaraldehyde: adrenaline and noradrenaline (Coupland & Hopwood, 1966) dopamine, dopa, 5-hydroxytryptamine, isoprenaline and β-phenylethylamine (Hopwood, 1967*c*, 1971*b*). In the case of the biogenic amines it was found, using spectroscopic techniques, that in the presence of excess glutaraldehyde the reactions were kinetically pseudo-first order. The reactions could lead to the formation of precipitates under high concentrations of primary amines at pH 7.2 (Hopwood, 1971*b*). This was used as the basis of an electron histochemical method to distinguish adrenaline- and noradrenaline-storing cells (Coupland *et al.*, 1966). Rost & Ewen (1971) have shown that glutaraldehyde vapour will react with various biogenic amines to give fluorescent compounds. The mechanism is similar to that described by Jonsson (1971) for formaldehyde and biogenic amine. In proteins, after reaction with glutaraldehyde, amino acid analyses have indicated that only the ε-amino group of lysine has reacted to any significant extent with glutaraldehyde (Tomimatsu *et al.*, 1971).

(b) PROTEINS

(i) *Enzymes.* There have now been a considerable number of studies on the reaction of glutaraldehyde with proteins showing changes in specific properties, e.g. immunological and enzyme activities, cross-linking, reaction kinetics, changes in conformation and staining properties.

As a general phenomenon Fortes (1971) pointed out that when glutaraldehyde reacts with a solution of an amino acid or protein the solution slowly becomes more acid. Presumably glutaraldehyde reacts with free amino groups. In turn this causes the following equilibrium to move over to the right.

$$R\text{-}NH_3^+ \rightleftharpoons R\text{-}NH_2 + H^+$$

Fixatives in general reduce enzyme activity. When Sabatini *et al.* (1963) introduced glutaraldehyde they made a semiquantitive estimate of the changes in activity for various enzymes. Mostly, a moderate activity remained. There have been numerous qualitative enzyme histochemical studies since then (Table 1). The methodology of fixation for this process have been reviewed recently by Shnitka & Sabatini (1971). In each case the enzyme activity was considerably reduced. Considerable inhibition of enzyme activity has been attributed to limitations in substrate diffusion and cross-link formation between the protein molecules (Quiocho *et al.*, 1964; Hopwood, 1969*b*). Arborgh *et al.* (1971) have looked at this problem again; they pointed out that previous studies were with blocks of tissue that had been immersed for long periods in the fixative. However, due to the slow rate of penetration of glutaraldehyde the complete fixation of all the tissue remained uncertain. They also studied the effect of perfusion glutaraldehyde fixation on the activities of acid phosphatase and aryl sulphatase in liver. After 5 min the activities of both enzymes, as determined by an assay, had fallen to 12% and 30% of the unfixed 'control' values. Washing of the perfused liver increased the enzyme activity. Inspection of Table 1 shows that when fixation of tissue blocks was relatively brief the enzyme activity remained high. After prolonged fixation, however, the remaining activity was at

Table 1. Inhibitions by glutaraldehyde of enzymes in various states.

Enzyme	*State*	*Reaction time*	*% Activity remaining*	*Temp. (°C)*	*% Glutaraldehyde (g/100 ml)*	*pH*	*Authors*
Acid phosphatase	Tissue blocks	1 hr	40	4	4	7.3	Anderson, 1967
		2 hr	28	4	4	7.2	Hopwood, 1967*a*
		6 hr	19–14	2	4	7.2	Janigan, 1965
		18 hr	15	4	4	7.2	Hopwood, 1967*a*
		24 hr	12	2	4	7.2	Janigan, 1965
	Perfused tissue	5 min	12	4–20	1.5	7.4	Arborgh *et al.*, 1971
N-acetyl-β-glucosaminidase	Tissue blocks	18 hr	54[a]	2	4	7.2	Janigan, 1964
Alanine aminotransferase	Tissue blocks	1 hr	50	4	4	7.3	Anderson, 1967
Aryl sulphatase	Perfused tissue	5 min	30	4–20	1.5	7.4	Arborgh *et al.*, 1971
Aspartate aminotransferase	Homogenate film	5 min	30–32	4	1	7.2	Papadimitrio *et al.*, 1970
ATPase, Na- & K-activated	Homogenate	40–60 min	0	4	0.5	7.3	Ernst *et al.*, 1970
ATPase, Mg-activated	Homogenate	40–60 min	15	4	0.5	7.3	Ernst *et al.*, 1970
Carboxypeptidase	Crystalline	1 hr	40	23	1	7.5	Quiocho *et al.*, 1966
Catalase	Crystalline	1 hr	12	23	4	7.2	Schejter *et al.*, 1970
	Tissue blocks	2 hr	12.5	4	4	7.2	Hopwood, 1967*a*
		18 hr	6.6	4	4	7.2	Hopwood, 1967*a*
Cholinesterase	Tissue blocks	1 hr	75	4	4	7.2	Anderson, 1967
α-Chymotrypsin	Crystalline	1 hr	0.4–1.2	0	2.3	6.2	Jansen *et al.*, 1971
iso-Citric dehydrogenase	Tissue blocks	1 hr	20	4	4	7.3	Anderson, 1967
Creatinine phosphokinase	Tissue blocks	1 hr	12	4	4	7.3	Anderson, 1967
β-Galactosidase	Tissue blocks	7 hr	43	2	4	7.2	Janigan, 1964
		24 hr	24	2	4	7.2	Janigan, 1964
β-Glucuronidase	Tissue blocks	2 hr	38	4	4	7.2	Hopwood, 1967*a*
		7 hr	24	4	4	7.2	Janigan, 1964
		18 hr	15	2	4	7.2	Hopwood, 1967*a*
		24 hr	12	4	4	7.2	Janigan, 1964
Glycogen phosphorylase *b*	Crystalline	10 min	40	23	0.05–0.01	7.5	Wang *et al.*, 1969
α-Hydroxybutyrate dehydrogenase	Tissue blocks	1 hr	30	4	4	7.3	Anderson, 1967
Lactate dehydrogenase	Tissue blocks	1 hr	13	4	4	7.3	Anderson, 1967
Ribonuclease	Crystalline		32	4	4	7.2	Sachs *et al.*, 1970
Subtilisin	Crystalline	30–60 min	10–15	23	2	6–8	Ogata *et al.*, 1968

Note: a = tissue washed in running water before assay.

the same level as that produced either by fixation, or by perfusion or after direct reaction between glutaraldehyde and purified enzyme.

Ottesen & Svensson (1971) found that when crystalline mercuripapain was reacted with various concentrations of glutaraldehyde (1–0.05 %) a similar amount of precipitate was formed, but at lower glutaraldehyde concentrations a higher specific protease activity was retained (1.5–51%). Associated with this was a decrease in the amount of lysine destroyed.

There is some evidence that when proteins are cross-linked in a monomolecular layer the product retains more activity than when it occurs in three dimensions (Walsh *et al.*, 1970; Axén, *et al.*, 1970).

Table I shows that increasing the time of fixation decreases enzyme activity. Arborgh *et al.* (1971) ascribe this as indicating slow penetration and lack of access of glutaraldehyde to the enzyme. Ottesen *et al.* (1971) have shown that enzyme activity also decreases with time of reaction in experiments in solution. From this, one can conclude that for enzyme histochemistry a short fixation period with a low concentration of glutaraldehyde would be optimal as far as preserving enzyme activity is concerned. This may also be a practical proposition as low concentrations of glutaraldehyde have been shown to give satisfactory ultrastructural preservation, e.g. 0.25 % (Maunsbach, 1966) and 0.4–0.5% (Hopwood, 1967*b*). A corrollary of these findings has been reported by Mantai (1970). He found that the rates of digestion of glutaraldehyde-fixed chloroplasts by lipase and trypsin was about 10–15% of the unfixed substrate. West & Mangan (1970) also found that the digestability of glutaraldehyde-treated kale chloroplasts by trypsin was reduced by 70%.

Papadimitriou & van Duijn (1970) investigated the principle of substrate protection for aspartate aminotransferase in the presence of formaldehyde and glutaraldehyde. With both fixatives the rate of inactivation was delayed if ketoglutarate was added to the fixative. The histochemical application of this principle should be investigated more widely.

(ii) *Reaction mechanisms.* Tomimatsu *et al.* (1971) have brought together evidence of the mechanism of enzyme inhibition by glutaraldehyde. Working with α-chymotrypsin they found that relatively small polymers had lost 60–70 % of their enzyme activity. They also pointed out that previous work had shown that acetylation of the ε-amino groups of lysine causes a considerable loss of enzyme activity (Janssen *et al.*, 1951). Wang & Tu (1969) also reported that modification of ε-amino groups of lysine in glycogen phosphorylase *b* by glutaraldehyde, dinitrophenol or potassium cyanate inactivated the enzyme completely or partly. Butyraldehyde, on the other hand, had no such effect. They concluded that some amino groups were essential for the functioning of the enzyme whereas others were not. Wang *et al.* (1969) made the important observation that glycogen phosphorylase *b* which had been modified by glutaraldehyde showed increased resistance to thermal denaturation.

After the reaction of proteins with glutaraldehyde, an amino acid analysis of the precipitate indicated that lysine is the only residue that was significantly changed, decreasing to about 50–60 % of the original content. The results of the analyses of a number of proteins and their glutaraldehyde reaction products are shown in Table II. Ottesen *et al.* (1971) found that a long reaction of mercuripapain with 2.3% glutaralde-

Table 2. Changes in lysine content, determined by amino acid analysis, of various proteins after reactions with glutaraldehyde.

Protein	*Lysine content*[a] *Original*	*Insolubilized*	% *decrease*	% *glutaraldehyde*	*Authors*
Bovine serum albumin	5.1	2.0	61	1.2	Jansen *et al.*, 1971
Carboxypeptidase A	15	6	60	1	Quiocho *et al.*, 1966
Carboxypeptidase A	15	4	63	4	
Catalase	103	56	46	4	Schejter *et al.*, 1971
α-Chymotrypsin	3.3	1.3	61	2.3	Jansen *et al.*, 1971
Chymotrypsinogen A	3.2	1.4	56	0.05	
Glycogen phosphorylase b			10	0.1	Wang *et al.*, 1969
β-Lactoglobulin			86	1	Bishop *et al.*, 1968
Lysozyme	2.3	1.0	57	0.025	Jansen *et al.*, 1971
Papain	2.3	1.0	57	0.1	
Soybean trypsin inhibitor	3.0	1.4	53	1.2	
Subtilisin	11	6.9	37	2	Ogata *et al.*, 1968
Wool (merino)	2.73	1.12	59	1	Marzona *et al.*, 1969

Note: a = in various units; for actual values see authors cited.

hyde gave an 83% loss of lysine. It would appear that this reaction is progressive with time, probably depending on the accessibility of the ε-amino groups. Some of these are essential for enzyme activity whereas others are not (for discussion, see Wang *et al.*, 1969). Marzona *et al.* (1969), Bowes & Cater (1968), Habeeb *et al.* (1968), Hopwood *et al.* (1970) and di Modica & Marzona (1971) have produced some evidence that aromatic amino acid residues in proteins may also react with glutaraldehyde. On the other hand, Tomimatsu *et al.* (1971) have denied this.

Hughes *et al.* (1970) reported that the ε-amino group of diaminopimelic acid in *B. subtilis* reacted with glutaraldehyde. There was a 17–27% reduction in this amino acid. This represented 30–50% of the available amino acid.

The ability of glutaraldehyde to form cross-links in tissues and between bovine serum albumin or casein and collagen molecules has been studied by Hopwood (1969*b*) and Bowes *et al.* (1968) using different techniques. Both Sephadex gel filtration and polyacrylamide gel electrophoresis can be used to separate molecules according to their mol. wt. Using this principle Hopwood (1969*b*) showed that glutaraldehyde was very efficient and rapid at cross-linking or polymerizing protein molecules compared with formaldehyde or a hydroxyadipaldehyde (Fig. 10).

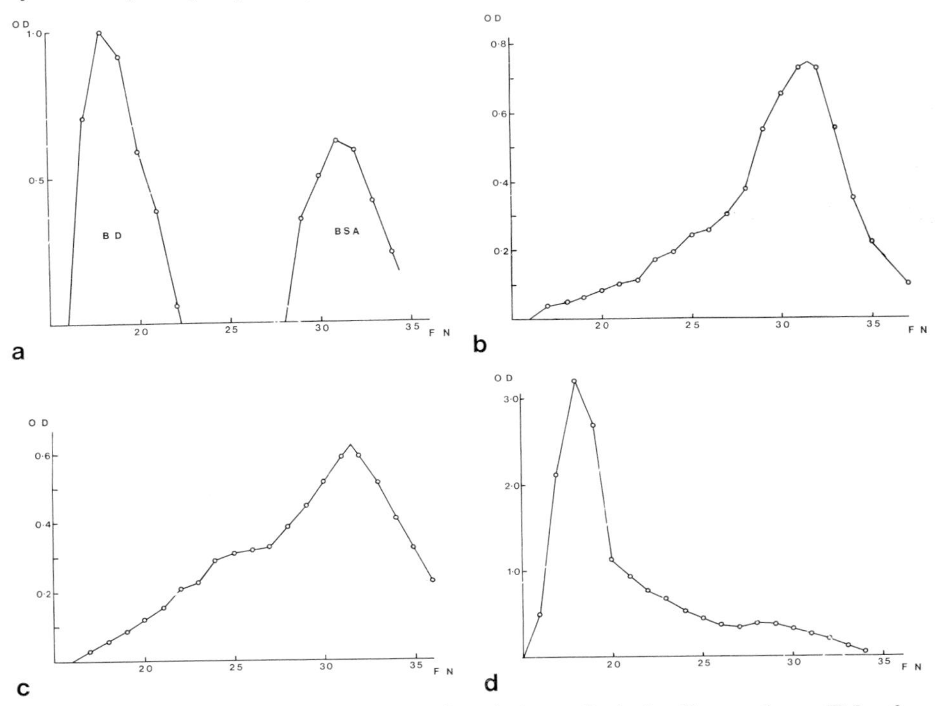

Figure 10. Elution patterns of bovine serum albumin from a Sephadex G-200 column. FN = fraction number (3 ml). OD = extinction at 280 nm: (*a*) Void volume is shown by blue dextran (BD) and elution curve of native bovine serum albumin (BSA); (*b*) elution curve of bovine serum albumin which had reacted with formaldehyde for 1 hr before gel filtration; (*c*) elution curve of bovine serum albumin treated with α-hydroxyadipaldehyde for 1 hr before gel filtration; (*d*) elution curve of bovine serum albumin which had reacted with glutaraldehyde for 1 hr before gel filtration. From Hopwood (1969*b*).

Ellar *et al.* (1971) compared fresh and glutaraldehyde-treated plasma membranes of *Micrococcus lysodeikticus* by polyacrylamide gel electrophoresis. They also found that glutaraldehyde treatment caused an increase of protein at the origin. This again must be due to heteropolymerization of the proteins which become too large to enter the pores of the gel. Bowes *et al.* (1965) using stress-strain measurements on rat tail collagen came to the same conclusion. Similar techniques (Hopwood, 1969) were used to investigate the intermolecular bond formation by osmium tetroxide, potassium dichromate and potassium permanganate with bovine serum albumin and γ-globulin. Osmium tetroxide was found to be the most efficient, the other two being relatively inefficient. Potassium permanganate also caused considerable oxidative cleavage of the protein molecules.

Hopwood (1969*c*) also showed that this was true in the case of tissue proteins. Thin slices (1 mm) of liver were fixed in formaldehyde, glutaraldehyde and hydroxyadipaldehyde for 1 hr and then inserted into slots of a starch gel slab and subjected to electrophoresis. The stained slab showed that no or very little protein had been pulled out of the glutaraldehyde-fixed material whereas the other fixed tissues lost as much protein as the unfixed control slices (Fig. 11). This difference in reactivity of proteins with the different aldehydes is also reflected in their modification of enzyme activity and antigenic properties (Hopwood, 1969*b*). This modification of antigenic properties has been used by Schechter (1971) to prolong the survival of glutaraldehyde-treated skin homografts in mice from 12 to 39 days. Donnelly & Goldstein (1970) also reported that after cross-linking with glutaraldehyde, concanavalin A still retains its ability to absorb polysaccharides and glycoproteins. Grillo *et al.* (1971) found that after the pancreas had been fixed in formaldehyde and sections prepared, biologically-active insulin could still be extracted.

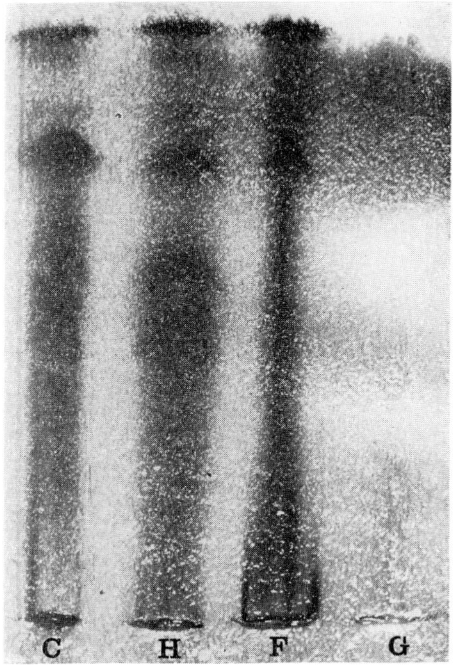

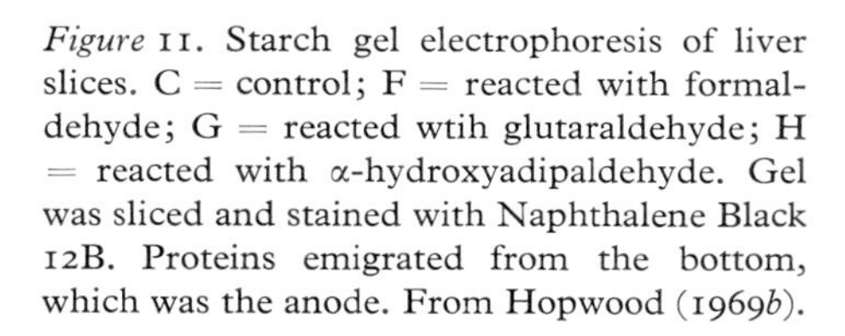

Figure 11. Starch gel electrophoresis of liver slices. C = control; F = reacted with formaldehyde; G = reacted wtih glutaraldehyde; H = reacted with α-hydroxyadipaldehyde. Gel was sliced and stained with Naphthalene Black 12B. Proteins emigrated from the bottom, which was the anode. From Hopwood (1969*b*).

The formation of cross-links in α-chymotrypsin and other proteins has been studied by Jansen *et al.* (1971) and Tomimatsu *et al.* (1971) using light-scattering measurements. They found that a two-step reaction was involved (Fig. 12). The initial rapid increase in light scattering was due to intermolecular cross-linking with especially reactive ε-amino groups of lysine residues which are on or near the surface of the molecule. The second increase in light scattering, they concluded, was due to linear polymerization in which the polymers formed in the first step crosslink to form large polymers. After 450 min reaction, polymer particles of mean mol. wt. 2.2×10^8 Daltons were formed, the particles being branched flexible coils. The product was not soluble in 6M urea. The reactions proceeded more rapidly with higher concentrations of glutaraldehyde. Loss of enzyme activity is associated with the formation of relatively small polymers, i.e. polymerization itself is not a cause of loss of activity. This would tie in well with the observations on rapid loss of enzyme activity. The implications of this finding and the later onset of a high degree of polymerization for fixation are obvious.

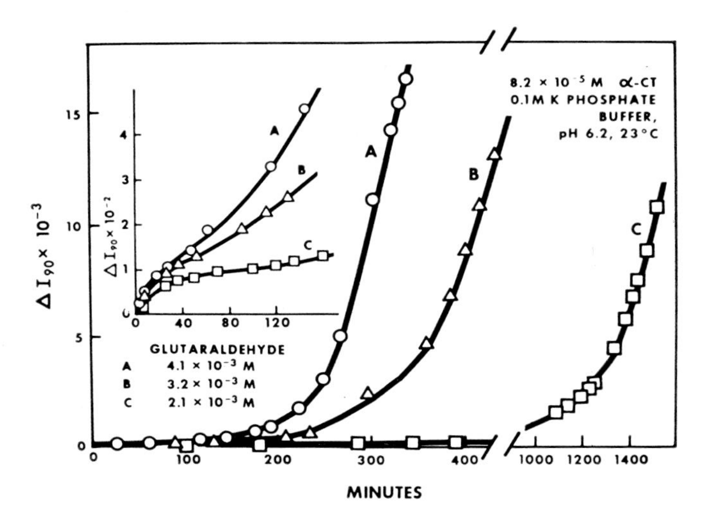

Figure 12. Change in 90° scattering (ΔI_{90}) with time as a function of glutaraldehyde concentration. From Tomimatsu *et al.* (1971) with the authors' permission.

Various workers have also investigated the effects of pH on the reaction of glutaraldehyde with proteins (Bowes & Cater, 1965; Habeeb *et al.*, 1968; Hopwood *et al.*, 1970; Sipes, 1970; Tomimatsu *et al.*, 1971). At lower pH levels, the reaction proceeds more slowly (Fig. 13). Tomimatsu *et al.* (1971) ascribed this to the stronger repulsive forces between the α-chymotrypsin molecules at low pH levels; over the pH range they studied, there would be relatively little increase in reactivity of the β-amino groups of lysine. The pH optimum for insolubilization of proteins was generally near their isoelectric point, although there were exceptions. Insolubilization was more rapid, they found, the lower the ionic strength of the reaction mixture. They also showed that when mixtures of proteins were made insoluble with glutaraldehyde the optimum results were obtained at a hydrogen ion concentration intermediate between that of the individual proteins. This suggests that the protein charge plays an important part in the process of protein insolubilization. The isoelectric point of most proteins is near neutrality although there are some exceptions (Young, 1963).

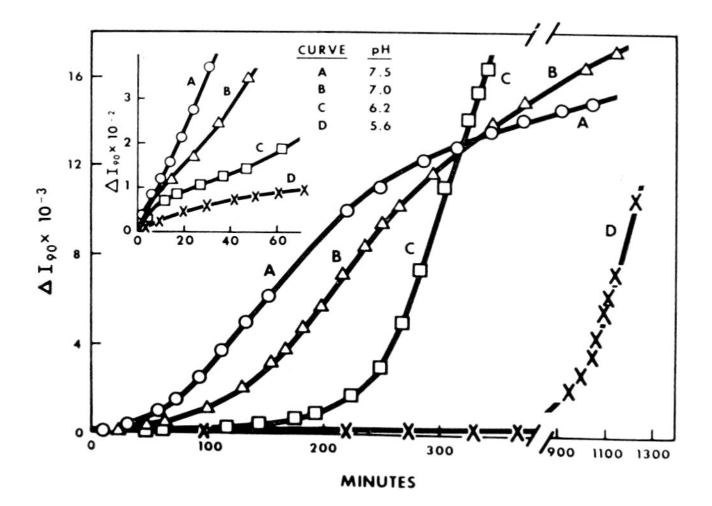

Figure 13. Effect of pH on the rate of change in 90° scattering ($\triangle I_{90}$). From Tomimatsu *et al.* (1971) with the authors' permission.

Tomimatsu *et al.* also showed that a stable soluble form of the protein-glutaraldehyde complex occurred. This was also found in polyacrylamide gel electrophoresis (Hopwood, 1969*b*) and in isoelectric focusing experiments (Hopwood, 1971*a*).

The chemical kinetics of the reactions between glutaraldehyde and a number of proteins have been studied by Hopwood *et al.* (1970). The protein-glutaraldehyde product had an absorption maximum at 275 nm. This has also been noted by Bowes *et al.* (1968) and Filacion *et al.* (1967). The nature of this new peak was investigated by Tomimatsu *et al.* (1971) who concluded it was due to increased light scattering, there being no significant change in tyrosine content. However, Hopwood *et al.* (1970) demonstrated spectroscopically that a reaction takes place between tyrosine and glutaraldehyde. A spectral shift to the blue was associated with the rapid appearance of a yellow colour in non-pigmented proteins. Some proteins also formed a precipitate with glutaraldehyde. This did not seem to be related to either their mol. wt. or their lysine or total aromatic amino acid content. The reaction between glutaraldehyde and the proteins was rapid and irreversible. The overall kinetics when glutaraldehyde is in excess are pseudo-first order. The temperature dependence of the reaction rates were also determined experimentally and found to obey the Arrhenius relationship

$$\log_{10} k = \text{constant} - E/2.303RT$$

where E is the activation energy. The results for the reaction with casein are shown in Fig. 14.

The activation energies were also determined for the reactions with bovine serum albumin and wheat acid phosphatase (Table 3). The values were interpreted as showing

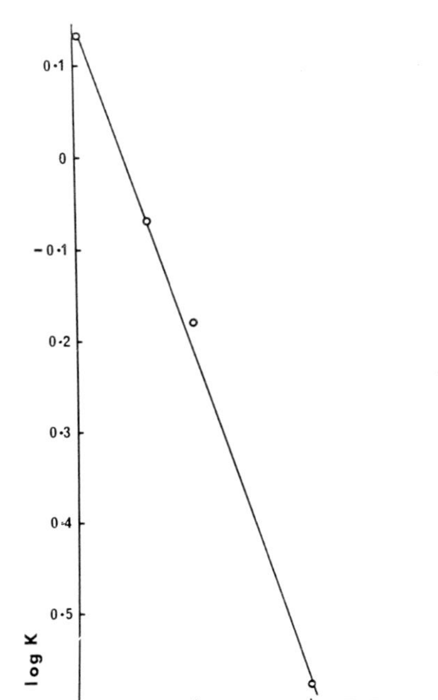

Figure 14. Determination of the activation energy for casein-glutaraldehyde at pH 7.2. From Hopwood *et al.* (1970).

that there is little denaturation of these proteins by glutaraldehyde as they are less than half those found in denaturation (Casey, 1962). Other work on changes in protein conformation has shown that other proteins, including myosin (Kaldor & Weinbach, 1965), chloroplast proteins (Park *et al.*, 1966), carboxypeptidase A (Quiocho *et al.*, 1964, 1966) and α-chymotrypsin (Tomimatsu *et al.*, 1971), are relatively little changed by glutaralde-

Table 3. Activation energies for the reaction between glutaraldehyde and various proteins compared with those of other processes of biological interest (from Hopwood *et al.*, 1970).

Process	*Activation energies*[a]
Glutaraldehyde + bovine serum albumin	10.6
Glutaraldehyde + casein	12.0
Glutaraldehyde + acid phosphatase	10.8
Molecule + molecule ⟶ products	
(hydrolyses, rearrangements)	
uncatalysed	15–30
catalysed	10–20
Denaturation of proteins and inactivation of enzymes	30–150[b]

Notes: a – units, kcal/mol; *b* – from Casey (1962).

hyde. Lenard & Singer (1968), using circular dichroism measurements in the spectral region of peptide absorption bands, found significant changes in the three dimensional conformation of intact red blood cell membranes, bovine albumin and sperm whale apomyglobin. The loss of helix ranged from 22 to 29%. Subsequent treatment of glutaraldehyde-fixed protein with osmium tetroxide or simply reaction of protein with potassium permanganate largely obliterated their helix content. Roux & Hillman (1969) have shown that glutaraldehyde changes the spectrum of oat and corn phytochrome in the red-absorbing region. This was interpreted as being due to protein conformational changes. At the tissue level these findings are in keeping with the results of some comparative fixation experiments by Sjöstrand & Barojas (1968) who investigated the effects of glutaraldehyde, osmium tetroxide and potassium permanganate.

The kinetics of the reaction of excess glutaraldehyde with red blood cells have been approached in another way by Morel *et al.* (1971). These workers followed the decrease in osmolarity of fixative solutions with time. There was an initial drop which they ascribed to the rapid entry of glutaraldehyde into the erythrocytes and the dilution of the fixative by the cell volume. Although the glutaraldehyde diffuses into the cells rapidly, it is not as rapid as water. This can lead to transient osmotic imbalances and changes in shape of cells and organelles, which are visualized with the electron microscope. The change in osmolarity appeared to fit equations like:

$$1/(\Omega - \Omega_1) = kt \times c$$

where Ω is the osmolarity at time t and Ω_1 the osmolarity at 5 hr, that is, the change is a pseudo-second order reaction, with respect to the glutaraldehyde and red blood cells. At later times the kinetics became more complex. Morel *et al.* also showed that the glutaraldehyde in the red blood cells was of two sorts. Some was in free solution in equilibrium with the external medium and the rest had reacted with the cell proteins. This last is equivalent to 120–140 molecules of glutaraldehyde per molecule of haemoglobin. They calculated that this could mean 100–300 possible sites for reaction, which agreed well with the known structure of haemoglobin.

(iii) *Changes in staining.* According to Chambers *et al.* (1968), when tissues have been fixed in glutaraldehyde, their staining properties are not the same as those of tissues which have been fixed in, for example, formaldehyde. These authors compared tissues from autopsy and surgical specimens fixed in glutaraldehyde with those fixed with formaldehyde. They found that glutaraldehyde-fixed tissues showed an increased affinity for acidic and basic dyes compared with the formaldehyde-fixed material. Workers in the textile industry, however, have found that glutaraldehyde treatment of wool slightly decreases its affinity for acid dyes (di Modica *et al.*, 1971). This effect of fixation on staining properties of tissues is well known (Lillie, 1967; Seki, 1936).

One phenomenon possibly related to this is the change in isoelectric point of proteins brought about by the action of fixatives. Hopwood (1971*b*) has shown that the technique of isoelectric focusing may be used to determine this. He showed that when bovine serum albumin was treated with formaldehyde or α-hydroxyadipaldehyde, osmium tetroxide or potassium dichromate there was relatively little change in the isoelectric point compared with the unfixed control material. However, glutaraldehyde produced two fractions of protein, the larger with a considerably lowered isoelectric point. Lowering of the isoelectric point had been noted previously in a semiquantitative manner (Tooze, 1964; Lillie, 1967) and quantitatively in the case of ε-chymotrypsin by Tomimatsu *et al.* (1971).

(iv) *Loss of materials.* Another phenomenon associated with the enhanced staining after glutaraldehyde fixation is related to the loss of cellular substance during fixation. The effectiveness of glutaraldehyde compared with other fixatives in preventing the loss of proteins from liver slices was mentioned earlier. The losses with various other fixatives have been investigated by Ostrowski *et al.* (1961) and Amsterdam & Schramm (1966). The effect of formaldehyde on the loss of small molecules from chromaffin tissue has been established on a quantitative basis by Hopwood (1967*c*, 1968). During fixation 30–40% of the catecholamines and ATP were lost into the fixative. This loss of material is in part due to the sensitivity of chromaffin granules to pH, as was shown by changes in the extinction at 520 nm. This technique was also used to compare the effects of formaldehyde and glutaraldehyde fixation on isolated ox chromaffin granules. Again, glutaraldehyde causes far less loss of material than formaldehyde (Fig. 15). Assay of the supernatant for proteins, nucleotides and catecholamines parallels the spectroscopic findings.

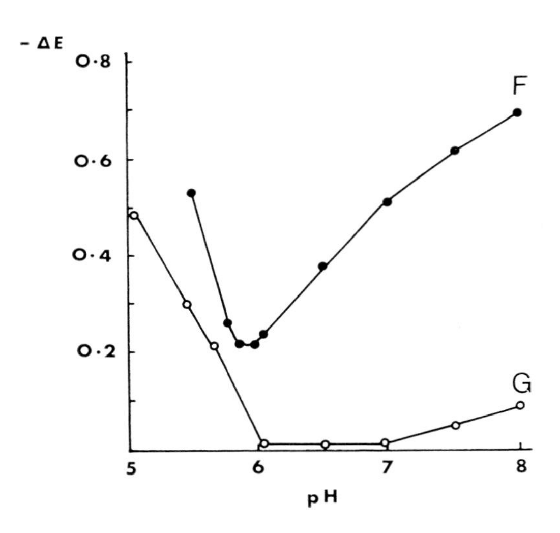

Figure 15. Effects of formaldehyde (F) and glutaraldehyde (G) on isolated ox chromaffin granules. Results expressed as changes ($\triangle E$) in extinction at 520 nm. Average results of six experiments. From Hopwood (1968).

Insulin is a fairly small protein molecule and Grillo *et al.* (1971) have compared the ability of glutaraldehyde, formaldehyde and ethanol to retain it in pancreas. There was some loss of insulin after immersion in any of these fixatives but the loss was least after glutaraldehyde fixation.

A further factor in the altered staining properties is the possibility that only one aldehyde group of some molecules of glutaraldehyde may react with the protein molecule, thus leaving the other aldehyde group free. Hopwood (1967*a*) investigated this problem with bovine serum albumin. After reaction with glutaraldehyde, the protein-aldehyde product was separated from the excess aldehyde on a Sephadex G-25 column. Bound aldehyde was then demonstrated biochemically as being associated with the protein.

The same phenomenon could be demonstrated in tissues, using Schiff's reagent to detect the free aldehyde groups. Morel *et al.* (1971) have come to the same conclusion.

Although there is now a considerable amount of information available on how glutaraldehyde reacts with proteins, the nature of the reaction still remains obscure. Ottesen *et al.* (1971) pointed out that depending on the temperature, pH and reagent concentration, glutaraldehyde and papain could react together in a number of ways to give rise to bonds of varying stability. Amino acid analyses have shown that the ε-amino groups of lysine are the main site of the reaction. Earlier workers suggested that a Schiff base was formed. The stability of the product and the kinetics of its formation (Richards *et al.*, 1968; Hopwood *et al.*, 1970) argue against this. Richards *et al.* (1968) have suggested the reaction involves α,β-unsaturated forms of glutaraldehyde. Bowes *et al.* (1968) have suggested that six glutaraldehyde molecules are involved in the reaction with the cross-linking of two lysine molecules. Future work should clarify these uncertainties.

(c) NUCLEIC ACIDS

Little work appears to have been done on the reaction of glutaraldehyde with nucleic acids. Formaldehyde is known to react with them under various circumstances (Berns & Thomas, 1961). It is possible to assess structural changes, if any, by measuring thermal transitions. Formaldehyde lowers the transition temperature. Hopwood (unpublished observations) has found that this reaction also occurs between glutaraldehyde and RNA of *Torula* and thymus DNA. The relevance of this lies in the curing of embedding resins at elevated temperatures which will cause the nucleic acids to allow any unreacted glutaraldehyde—whole or part molecules—to react with the exposed bases. When the material cools, the nucleic acids will remain distorted. Any electron microscopic work on nucleic acids using glutaraldehyde fixation must take this into account. Similar results have been obtained by Brooks & Klamerth (1968) who investigated the reactions of calf thymus DNA with glyoxal and malondialdehyde. They also found that the temperature-extinction profiles of reacted DNA were considerably changed compared with native DNA. They also found that dialdehyde-treated DNA showed considerable resistance to deoxyribonuclease. These workers concluded that the dialdehydes had reacted with the amino groups of cytidine and guanine.

Ellar *et al.* (1971) showed that glutaraldehyde treatment of *M. lysodeikticus* caused only a slight binding of nucleic acids to plasma membranes, although there was a considerable formation of intermolecular crosslinks between proteins.

(d) LIPIDS

The reactions between lipids and glutaraldehyde have been little explored. In 1965 Levy *et al.* reported that glutaraldehyde had little effect on the extraction of brain lipids compared with fresh, unfixed tissue controls. Roozemond (1969) compared the fixation of phospholipids in the rat hypothalamus with formaldehyde and glutaraldehyde. He found that glutaraldehyde was more efficient in preventing the loss of phosphatidylserine and phosphatidylethanolamine. He concluded that this was due to the fixing of these phospholipids through their amino groups to the tissue proteins. L'Hermite & Israel (1969) found that glutaraldehyde fixation enhanced the autofluorescence of rat myelin and certain perikaryal inclusions. This was reinforced if calcium chloride was present in the fixative solution. They isolated the myelin by centrifugation, separated its components by t.l.c. and investigated the effect of glutaraldehyde on the individual components. There was an enhancement of the autofluorescence of cerebrosides, phosphatides and proteolipids. The reactivity of glutaraldehyde with these lipids is probably through their nitrogen atom as Roozemond (1969) suggested.

The ability of aldehydes to retain lipids has been noted by Ljubešić (1970) working with blackberry (*Rubus fructicosus*) where he found some cells and their plastids subsequently very osmiophilic.

(e) MUCOSUBSTANCES

Hopwood (1967*a*) compared the preservation of glycogen in rat liver by formaldehyde and glutaraldehyde. The percentages retained were 75 and 65 respectively. The best preservative of glycogen is cold ethanol as Kugler & Wilkinson have demonstrated. Cejkova & Brettschneider (1970) have reported that 24 hr fixation of rabbit cornea by glutaraldehyde had no effect on the acid mucopolysaccharides assayed biochemically. Evans (1969) pointed out that there were many reports in the patent literature of the reaction between glutaraldehyde and polyhydroxyl compounds, e.g. pentaerythritol, to produce polymers. Such reactions may take place with mucosubstances thereby cross-linking them.

$$HOCH_2-C(CH_2OH)_3 \longrightarrow \left[-CH{<}^{O.CH_2}_{O.CH_2}{>}C{<}^{CH_2.O}_{CH_2.O}{>}CH.CH_2.CH_2.CH_2- \right]_n$$

Conclusions

Since its introduction as a fixative for electron microscopy, the use of glutaraldehyde has become widespread. The fine structure found is reproducible and detailed. Something of the mechanisms involved is now understood in respect of proteins in general and enzymes in particular. Changes in tertiary structure are now being described. Little is known of the reactions of glutaraldehyde with nucleic acids, mucosubstances and lipids. This hiatus in our knowledge will no doubt be filled in over the next few years. Our poor understanding of fixation is in many ways rather surprising as a huge system of knowledge has been built up which depends on this ill-comprehended step. This applies both to normal and pathological tissues. This gap must be filled as the morphological and biochemical sciences are converging. High resolution electron microscopy can now reveal various macromolecules. It is important to be able to interpret these findings in the light of biochemical descriptions for them.

Acknowledgements

I wish to thank Miss Colleen Peel for her cheerful typing of this review. I would also thank the following publishers for permission to reproduce material whose copyright they hold: Academic Press, Figs. 1*a*, *b*, *c*, 12 & 13; Chapman and Hall, Fig. 14 and Table 3; Cambridge University Press, Fig. 15; Elsevier, Figs. 5, 6 & 7; North Holland, Figs. 8 & 9; Springer-Verlag, Figs. 2, 3, 4, 10 & 11.

References

ALEXA, G., CHISALITA, D. & CHIRITA, G. (1971). Reaction of dialdehyde with functional groups in collagen. *Rev. Tech. Ind. Cuir.* **63,** 5–6.

AMSTERDAM, A. & SCHRAMM, M. (1966). Rapid release of zymogen granule protein by osmium tetroxide and its retention by glutaraldehyde. *J. Cell Biol.* **29,** 199–207.

ANDERSON, P. J. (1967). Purification and quantitation of glutaraldehyde and its effects on several enzyme activities in skeletal muscle. *J. Histochem. Cytochem.* **15,** 652–61.

ARBORGH, B., ERICSSON, J. L. E. & HELMINEN, H. (1971). Inhibition of renal acid phosphatase and arylsulphatase activity by glutaraldehyde fixation. *J. Histochem. Cytochem.* **19,** 449–51.

ASO, C. & AITO, Y. (1962). Studies on the polymerization of bifunctional monomers. II polymerization of glutaraldehyde. *Makromol. Chem.* **58,** 195–203.

AXÉN, R., MYRIN, P. Å. & JANSSON, J. C. (1970). Chemical fixation of chymotrypsin to water insolubilized cross linked Dextran (Sephadex) and solubilization of the enzyme derivatives by means of a dextranase. *Biopolymers* **9,** 401–13.

BAHR, G. F. (1954). Osmium tetroxide and ruthenium tetroxide and their reactions with biologically important substances. *Exp. Cell Res.* **7,** 457–89.

BAKER, J. R. (1960). *Principles of Biological Microtechnique.* London: Methuen.

BEAUVILLAIN, J. C. (1970). Influence de divers fixateurs sur la structure fine de L'éminence médiane de Souris. *C.R. Soc. Biol.* **164,** 1032–34.

BERNS, K. I. & THOMAS, C. A. (1961). A study of single polynucleotide chains derived from T2 and T4 bacteriophage. *J. Mol. Biol.* **3,** 289–300.

BISHOP, W. H. & RICHARD, F. M. (1968). Isoelectric point of a protein in the cross linked state: -lactoglobulin. *J. Mol. Biol.* **33,** 415–21.

BLOEM, E. (1968). Thin layer chromatography of aldehydes in leather. *J. Soc. leath. Trades Chem.* **52,** 204–7.

BLUMENFELD, O. O., PAZ, M. A., GALLOP, P. M. & SEIFTER, S. (1963). The nature, quantity and mode of attachment of hexoses in ichthyocol. *J. biol. Chem.* **238,** 3835–9.

BODIAN, D. (1970). An electron microscopic characterization of synaptic vesicles by means of controlled aldehyde fixation. *J. Cell Biol.* **44,** 115–24.

BOHMAN, S. O. & MAUNSBACH, A. B. (1970). Effects on tissue fine structure of variations in colloid osmotic pressure of glutaraldehyde fixatives. *J. Ultrastruct. Res.* **30,** 195–208.

BONE, Q. & DENTON, E. J. (1971). The osmotic effects of electron microscope fixatives. *J. Cell Biol.* **49,** 571–81.

BONE, Q. & RYAN, K. P. (1972). Osmolarity of osmium tetroxide and glutaraldehyde fixatives. *Histochem. J.* **4,** 331–47.

BOWES, J. A. & CATER, C. W. (1965). Cross linking of collagen. *J. Appl. Chem.* (London) **15,** 296–304.

BOWES, J. A. & CATER, C. W. (1968). The interaction of aldehydes with collagen. *Biochem biophys. Acta.* **168,** 341–52.

BROOKS, B. R. & KLAMERTH, O. L. (1968). Interaction of DNA with bifunctional aldehydes. *Europ. J. Biochem.* **5,** 178–82.

CARSTENSEN, E. L., ALDRIDGE, W. G., CHILD, S. Z., SULLIVAN, P. & BROWN, H. H. (1971). Stability of cells fixed with glutaraldehyde and acrolein. *J. Cell Biol.* **50,** 529–32.

CASEY, E. J. (1962). *Biophysics: Concepts and Mechanisms.* New York: Reingold.

CEJKOVA, J. & BRETTSCHNEIDER, I. (1970). Glutaraldehyde fixation of corneal acid mucopolysaccharides. *Ophthalmic Res.* **1,** 149–55.

CHAMBERS, R. W., BOWLING, M. C. & GRIMLEY, P. M. (1968). Glutaraldehyde fixation in routine histopathology. *Archs. Path.* **85,** 18–30.

CHIANG, J., KOWADA, M., AMES, A., WRIGHT, R. L. & MAJNO, G. (1968). Cerebral ischaemia III vascular changes. *Am. J. Path.* **52,** 455–65.

CHISALITA, D., CHIRITA, G. & ALEXA, G. (1971). Kinetics of dialdehyde combination with the reactive groups of collagen. *Ind. Usoara* **18,** 269–81 (Roum).

CLARK, M. A. & ACKERMAN, G. A. (1971). Alteration of nuclear and nucleolar pyroantimonate-osmium reactivity by glutaraldehyde fixation. *J. Histochem. Cytochem.* **19,** 388–90.

CLELAND, W. W. (1967). In: *Biological Chemistry* (eds. H. R. Mahler & E. H. Cordes), p. 224. London: Harper & Row.

CLIFT, F. P. & COOK, R. P. (1932). A method of determination of some biologically important aldehydes and ketones with special reference to pyruvic acid and methyl glyoxal. *Biochem. J.* **26,** 1788–99.

COUPLAND, R. E. & HOPWOOD, D. (1966). The mechanism of the differential staining reaction for adrenaline and noradrenaline storing granules in tissues fixed in glutaraldehyde. *J. Anat.* **100,** 227–43.

CURGY, J. J. (1968). Influence du mode de fixation sur la possibilité d'observer des structures myéliniques dans les hépatocytes d'embryons de poulet. *J. Microscopie* **7,** 63–80.

DANEEL, S. & WEISSENFELS, N. (1965). Besseres Fixierungsverfahren zur Darstellung des Grundplasmas von Protozoen und Vertebraten zellen. *Mikroskopie* **20,** 162–4.

DEAMER, D. W. & CROFTS, A. (1967). Action of Triton X-100 on chloroplast membranes. *J. Cell Biol.* **33,** 395–410.

DE JONG, D. W., OLSON, A. C. & JANSEN, E. F. (1967). Glutaraldehyde activation of nuclear acid phosphatase in cultured plant cells. *Science* **155,** 1672–4.

DESCARRIES, L. & DROZ, B. (1970). Intraneural distribution of exogenous norepinephine in the central nervous system of the rat. *J. Cell Biol.* **44,** 385–99.

di MODICA, G. & MARZONA, M. (1971). Cross linking of wool keratin by bifunctional aldehydes, *Textile Res. J.* **41,** 701–5.

DONNELLY, E. H. & GOLDSTEIN, I. J. (1970). Glutaraldehyde insolubilized Concanavalin A: an adsorbent for the specific isolation of polysaccharides and glycoproteins. *Biochem. J.* **118,** 679–80.

ELLAR, D. J., MUNOZ, E. & SALTON, M. R. (1971). The effect of low concentrations of glutaraldehyde on *Micrococcus lysodeikticus* membranes: changes in the release of membrane associated enzymes and membrane structure. *Biochem. biophys. Acta* **225,** 140–50.

ERICSSON, J. L. E. & BIBERFELD, P. (1967). Studies on aldehyde fixation. Fixation rates and their relation to fine structure and some histochemical reactions in liver. *Lab. Invest.* **17,** 281–98.

ERNST, S. A. & PHILPOTT, C. W. (1970). Preservation of Na-K-activated and Mg-activated ATP-ase activities of avian salt gland and teleost gill with formaldehyde as fixative. *J. Histochem. Cytochem.* **18,** 251–63.

EVANS, A. P. (1969). Reactions of glutaraldehyde and succinaldehyde with compounds containing replaceable hydrogen atoms. *Univ. Microfilms Inc. Ann Arbor Michigan* **70–11,** 634.

FAHIMI, H. D. & DROCHMANS (1965*a*). Essais de standardization de la fixation au glutaraldéhyde I. Purification et détermination de la concentration du glutaraldehyde. *J. Microscopie* **4,** 725–36.

FAHIMI & DROCHMANS, P. (1965*b*). Essais de standardization de la fixation au glutaraldéhyde II. Influence des concentrations en aldéhyde et de l'osmolalité, *J. Microscopie* **4,** 737–48.

FEIN, M. L. & HARRIS, E. H. (1962). Quantitative analytical procedure for determining glutaraldehyde and chrome in tanning solutions. U.S. Dept. Agriculture Publication ARS-73-37.

FILACHIONE, E. M., KORN, A. H. & ARD, J. S. (1967). The ultraviolet absorption of protein bound glutaraldehyde. *J. Am. Leather Chemists Assoc.* **67,** 450–3.

FLITNEY, E. W. (1966). The time course of fixation of albumin by formaldehyde, glutaraldehyde, acrolein and other higher aldehydes. *J.R. microsc. Soc.* **85,** 353–64.

FORSSMANN, W. G., SIEGRIST, G., ORCI, L., GIRARDIER, L., PICTET, R., ROUILLER, C., BAUMANN, M. & MORITZ, A. (1967). Fixation par perfusion pour la microscopie electronique: essai de generalisation. *J. Microscopie* **6,** 279–304.

FORTES, P. A. (1971). Glutaraldehyde as a probe of metabolism induced changes in mitochondrial proteins. *Proc. Colloq. Johnson Res. Found.*

FRANKE, W. W., KREIN, S. & BROWN, R. M. (1969). Simultaneous glutaraldehyde-osmium tetroxide fixation with post osmication. *Histochemie* **19,** 162–4.

FRENCH, D. & EDSALL, J. T. (1945). The reaction of formaldehyde with amino acids and proteins. *Adv. prot. Chem.* **2,** 277–335.

FRIGERIO, N. A. & SHAW, M. J. (1969). A simple method for determination of glutaraldehyde. *J. Histochem. Cytochem.* **17,** 176–81.

GARRETT, J. R., DAVIES, K. J. & PARSONS, P. A. (1972). Consumers' guide to glutaraldehyde. *Proc. R. microsc. Soc.* **7**, 116–17.

GIL, J. & WEIBEL, E. R. (1971). An improved apparatus for perfusion. *J. Microsc.* **94**, 241–4.

GRANTHAM, J., CUPPAGE, F. E. & FANESTIL (1971). Direct observation of toad bladder response to vasopressin. *J. Cell Biol.* **48**, 695–9.

GRILLO, T. A. I., OGUNNAIRE, P. O. & FAOYE, S. (1971). Effects of histological and electron microscopical fixatives on the insulin content of rat pancreas. *J. Endocr.* **51**, 645–9.

HAAS, D. J. (1968). Preliminary studies on the denaturation of crosslinked lysozyme crystals. *Biophysical J.* **8**, 549–55.

HABEEB, A. F. S. A. & HIRAMOTO, R. (1968). Reaction of proteins with glutaraldehyde. *Arch. Biochem. Biophys.* **126**, 16–26.

HANKER, J. S., KASLER, F., BLOOM, M. G., COPELAND, J. S. & SELIGMAN, A. M. (1967). Coordination of polymers of osmium: the nature of osmium black. *Science* **156**, 1737–8.

HARDIN, J. H. & SPICER, S. S. (1970). Ultrastructure of neuronal nucleoli of rat trigeminal ganglia. Comparison of routine with pyroantimonate-osmium tetroxide fixation. *J. Ultrastruct. Res.* **31**, 16–36.

HARDY, P. M., NICHOLLS, A. C. & RYDON, H. N. (1969). Nature of glutaraldehyde in aqueous solution. *Chem. Commun.* 565–6.

HARRIES, C. & TANK, L. (1908). Ueber die Aufspaltung der Cyclopentens zum Halbaldehyd der Glutarsaüre bezw. zum Glutardialdehyd. *Berlin Ber D. Chem. Ges.* **41**, 1701–11.

HAYAT, M. A. (1970). *Principles and Techniques of Electron Microscopy. Biological applications.* London: Van Nostrand Reinhold Co.

HECKER, H. (1970). Ultrastruktur der Symbioten in Ovozyten von *Ornithodorus moubata*, Murray (Ixodoidea: Argasidae) nach simultaner Glutaraldehyd-Osmiumfixierung und Nach behandlung mit Uranylacetat (Triple Fixation). *Experimentia* **26**, 874–7.

HELLER, J., OSTWALD, T. J. & BOK, D. (1971). The osmotic behaviour of rod photoreceptor outer segment discs. *J. Cell Biol.* **48**, 633–49.

HIRSCH, J. G. & FEDORKO, M. E. (1968). Ultrastructure of human leukocytes after simultaneous fixation with glutaraldehyde and osmium tetroxide and 'post fixation' in uranyl acetate. *J. Cell Biol.* **38**, 615–27.

HODSON, S. & MARSHALL, J. (1967). Tyrosine incorporation into the rabbit retina. *J. Cell Biol.* **35**, 722–6.

HOPWOOD, D. (1967*a*). Some aspects of fixation with glutaraldehyde. A biochemical and histochemical comparison of the effects of formaldehyde and glutaraldehyde fixation on various enzymes and glycogen with a note on penetration of glutaraldehyde into liver. *J. Anat.* **101**, 83–92.

HOPWOOD, D. (1967*b*). The behaviour of various glutaraldehydes on Sephadex G-10 and some implications for fixation. *Histochemie* **11**, 289–95.

HOPWOOD, D. (1967*c*). The effect of formaldehyde fixation and dehydration on ox adrenal medulla with respect to the chromaffin reaction and post-chroming. *Histochemie* **10**, 98–106.

HOPWOOD, D. (1968). The effect of pH and various fixatives on isolated ox chromaffin granules with respect to the chromaffin reaction. *J. Anat.* (Lond.) **102**, 415–24.

HOPWOOD, D. (1969*a*). Fixatives and fixation: a review. *Histochem. J.* **1**, 323–60.

HOPWOOD, D. (1969*b*). A comparison of the crosslinking abilities of glutaraldehyde, formaldehyde and α-hydroxydipaldehyde with bovine serum albumin and casein. *Histochemie* **17**, 151–61.

HOPWOOD, D. (1969*c*). Fixation of proteins by osmium tetroxide, potassium dichromate and potassium permanganate. *Histochemie* **18**, 250–60.

HOPWOOD, D. (1970). The reactions between formaldehyde, glutaraldehyde and osmium tetroxide and their fixation effects on bovine serum albumin and on tissue blocks. *Histochemie* **24**, 56–64.

HOPWOOD, D. (1971*a*). The histochemistry and electron histochemistry of chromaffin tissue. *Progr. Histochem. Cytochem.* **3**, 1–66.

HOPWOOD, D. (1971*b*). Use of isoelectric focusing to determine the isoelectric point of bovine serum albumin after treatment with various common fixatives. *Histochem. J.* **3**, 201–5.

HOPWOOD, D. (1972*a*). Fixation with mercury compounds. *Progr. Histochem. Cytochem.* **4**, 193–224.

HOPWOOD, D. ALLEN, C. R. & MCCABE, M. (1970). The reactions between glutaraldehyde and various proteins. An investigation of their kinetics. *Histochem. J.* **2,** 137–50.

HORTON, A. A. & PACKER, L. (1969). Mitochondrial metabolism of aldehydes. *Biochem. J.* **116,** 19–20.

HORTON, A. A. & PACKER, L. (1969*a*). Mitochondrial metabolism of aldehydes. *Biochem. J.* **116,** *19P–20P.*

HORTON, A. A. & PACKER, L. (1970). Interactions between malondialdehyde and rat liver mitochondria. *J. Gerentol* **25,** 199–204.

HUGHES, R. C. & THURMAN, P. F. (1970). Crosslinking of bacterial cell walls with glutaraldehyde. *Biochem. J.* **119,** 925–6.

JAHN, K. (1971). Über histochemische Aldehydreaktionen nach Fixation mit Glutaraldehyd. *Acta Histochem.* **39,** 298–301.

JANIGAN, D. T. (1964). Tissue enzyme fixation studies. *Lab. Invest.* **13,** 1038–50.

JANIGAN, D. T. (1965). The effect of aldehyde fixation on acid phosphatase activity in tissue blocks. *J. Histochem. Cytochem.* **13,** 473–83.

JANSEN, E. F. & OLSON, A. C. (1969). Properties and enzymatic activities of papain insolubilized with glutaraldehyde. *Arch. Biochem. Biophys.* **129,** 221–7.

JANSEN, E. F., TOMIMATSU, Y. & OLSEN, A. C. (1971). Cross linking of α-chymotrypsin and other proteins by reaction with glutaraldehyde. *Arch. Biochem. Biophys.* **144,** 394–400.

JANSEN, E. G., CURL, AL. & BALLS, A. K. (1951). A crystalline, active oxidation product of α-chymotrypsin. *J. Biol. Chem.* **189,** 671–82.

JARD, S., BOURGET, J., CARASSO, N. & FAVARD, P. (1966). Action des fixatives sen la perméabilité et l'ultrastructure de la véssie de grenouille. *J. Microscopie* **5,** 31–50.

JONSSON, G. (1971). Quantitation of fluorescence of biogenic amines. *Progr. Histochem. Cytochem.* **2,** 299–334.

KALDOR, G. & WEINBACH, S. (1965). Use of toluidine diisocyanide and glutaric dialdehyde in the conjugation of heavy meromyosin with light meromyosin. *Arch. Biochem. Biophys.* **112,** 448–52.

KANAMURA, S. (1970). Difference in resistance to glutaraldehyde or formaldehyde fixation between mouse and rat hepatic glucose 6-phosphatase. *Acta Histochem. Cytochem.* **3,** 160–2.

KANAMURA, S. (1971). Demonstration of glucose 6-phosphatase activity in hepatocytes following transparenchymal perfusion fixation with glutaraldehyde. *J. Histochem. Cytochem.* **19,** 386–7, 520–1.

KARNOWSKY, M. J. (1967). The ultrastructural basis of capillary permeability. Studies with peroxidase as a trace. *J. Cell Biol.* **36,** 213.

KAY, D. (1967). *Techniques for Electron Microscopy.* 2nd edn. Oxford: Blackwell.

LANDON, D. N. (1970). The influence of fixation upon the fine structure of the Z-disc of the rat striated muscle. *J. Cell Sci.* **6,** 257–76.

LENARD, J. & SINGER, S. J. (1968). Alterations of the conformation of proteins in red blood cell membranes and in solution by fixatives used in electron microscopy. *J. Cell Biol.* **37,** 117–21.

LEVY, W. A., HERZOG, I., SUZUKI, K., KATZMAN, R. & SCHEINBERG, L. (1965). Method for combined ultrastructural and biochemical analysis of normal tissue. *J. Cell Biol.* **27,** 119–32.

L'HERMITE, P. & ISRAEL, M. (1969). Action du glutaraldéhyde sur les lipide et les protéines myéliniques. *Ann. Histochim.* **14,** 1–11.

LILLIE, R. D. (1967). Hot chromation oxyphilia. *Histochemie* **11,** 332–49.

LJUBEŠIĆ, N. (1970). Osmiophile substanz in Blattzellen der Brombeere (*Rubus fructicosus* L.S.L.). *Protoplasma* **69,** 49–59.

MANTAI, K. E. (1970). Some effects of hydrolytic enzymes on coupled and uncoupled electron flow in chloroplasts. *Plant Physiol.* **45,** 563–6.

MARZONA, M. & MODICA, G. (1969). Stabilization of wool by difunctional aldehydes. *Ann. Chim. (Rome)* **59,** 956–62.

MAUNSBACH, A. B. (1966). The influence of different fixatives and fixation methods on the ultrastructure of rat kidney proximal tubule cells. *J. Ultrastruct. Res.* **15,** 283–309.

MILCH, R. A. (1963). Studies on collagen tissue aging: interaction of certain intermediary metabolites with collagen. *Gerontologia* **7,** 129–52.

MOLLENHAUER, H. H. & TOTTEN, C. (1971). Studies on seeds 1. Fixation of seeds. *J. Cell Biol.* **48,** 387–94.

MOREL, F. M. M., BAKER, R. F. & WAYLAND, H. (1971). Quantitation of human red blood cell fixation by glutaraldehyde. *J. Cell Biol.* **48,** 91–100.

MUNTON, T. J. & RUSSELL, A. D. (1970). Aspects of the action of glutaraldehyde on *Escherichia coli*. *J. appl. Bact.* **33,** 410–19.

OGATA, K., OTTESEN, M. & SVENDSEN, I. (1968). Preparation of water-insoluble, enzymically active derivatives of subtilisin type Novo by cross linking with glutaraldehyde. *Biochem. biophys. Acta* **159,** 403–5.

OSTROWSKI, K., KOMENDER, J. & KWARECKI, K. (1961). Quantitative investigations of the solubility of proteins extracted from tissues fixed by different chemical and physical methods. *Ann. Histochem.* **6,** 501–6.

OTTESEN, M. & SVENSSEN, B. (1971). Modification of papian by treatment with glutaraldehyde under reducing and non-reducing conditions. *C.R. Lab. Carlsberg* **38,** 171–85.

PACKER, L., ALLEN, J. M. & STARKS, M. (1968*a*). Light induced transport in glutaraldehyde fixed chloroplasts: studies with nigercin. *Arch. Biochem.* **128,** 142–52.

PACKER, L. & GREVILLE, G. D. (1969). Energy linked oxidation of glutaraldehyde by rat liver mitochondria. *FEBS Lett* **3,** 112–14.

PACKER, L., WRIGGLESWORTH, J. M., FORTES, P. A. G. & PRESSMAN, B. C. (1968*b*). Expansion of the inner membrane compartment and its relation to mitochondrial volume-ion transport. *J. Cell Biol.* **39,** 382.

PAPADIMITRIOU, J. M. & VAN DUIJN, P. (1970). Effects of fixation and substrate protection on the isoenzymes of aspartate aminotransferase studied in a quantitative cytochemical model system. *J. Cell Biol.* **47,** 71–83.

PARK, R. B., KELLY, J., DRURY, S. & SAUER, K. (1966). The Hill reaction of chloroplasts isolated from glutaraldehyde-fixed spinach leaves. *Proc. Nat. Acad. Sci. U.S.A.* **55,** 1056–62.

PEARSE, A. G. E. (1968). *Histochemistry: Theoretical and Applied*, 3rd edn., vol. 1. London: Churchill.

PETERS, T. & ASHLEY, C. A. (1967). An artefact in autoradiography due to binding of free amino acids to tissues by fixatives. *J. Cell Biol.* **33,** 53–60.

QUIOCHO, F. A. & RICHARDS, F. M. (1964). Intermolecular crosslinking of a protein in the crystalline state: carboxypeptidase-A. *Proc. Nat. Acad. Sci. U.S.A.* **52,** 833–8.

QUIOCHO, F. A. & RICHARDS, F. M. (1966). The enzymic behaviour of carboxypeptidase-A in the solid state. *Biochem.* **5,** 4062–76.

RABINOVITCH, M. & GARY, P. P. (1968). Effect of the uptake of staphylococci on the ingestion of glutaraldehyde treated red cells attached to macrophages. *Expt. Cell Res.* **52,** 363–9.

REALE, E. & LUCIANO, L. (1970). Fixierung mit Aldehyden. Ihre Eignung für histologische und histochemische Untersuchungen in der Licht- und Elecktronen-mikroskopie. *Histochemie* **23,** 144–70.

RICHARDS, F. M. & KNOWLES, J. R. (1968). Glutaraldehyde as a protein cross linking agent. *J. Mol. Biol.* **37,** 231–3.

ROBERTSON, E. A. & SCHULTZ, R. L. (1970). The impurities in glutaraldehyde and their effect on the fixation of the brain. *J. Ultrastruct. Res.* **30,** 275–87.

ROOZEMOND, R. C. (1969). The effect of fixation with formaldehyde and glutaraldehyde on the composition of phospholipids extractable from rat hypthalamus. *J. Histochem. Cytochem.* **17,** 482–6.

ROST, F. W. D. & EWEN, S. W. B. (1971). New methods for the histochemical demonstration of catechol amines, tryptamines, histamine and other arylethylamines by acid- and aldehyde-induced fluorescence. *Histochem. J.* **3,** 207–12.

ROUX, S. J. & HILLMAN, W. S. (1969). The effect of glutaraldehyde and two monoaldehydes on phytochrome. *Arch. Biochem. Biophys.* **131,** 423–9.

RUBBO, S. D., GARDNER, J. F. & WEBB, R. L. (1967). Biocidal activities of glutaraldehyde and related compounds. *J. appl. Bact.* **30,** 78–87.

SABATINI, D. D., BENSCH, K. & BARNNETT, R. J. (1963). Cytochemistry and electron microscopy. The preservation of cellular ultrastructure and enzymatic activity by aldehyde fixation. *J. Cell Biol.* **17,** 19–58.

SACHS, D. H. & WINN, H. J. (1970). The use of glutaraldehyde as a coupling agent for ribonuclease and bovine serum albumin. *Immunochem.* **7,** 581–5.

SCHECHTER, I. (1971). Prolonged survival of glutaraldehyde treated skin homografts. *Proc. Nat. Acad. Sci. U.S.A.* **68,** 1590–3.

SCHEJTER, A. & BARELI, A. (1970). Preparation and properties of cross linked water-insoluble catalase. *Arch. Biochem. Biophys.* **136,** 325–30.

SCHULTZ, R. L. & KARLSSON, U. (1965). Fixation of the central nervous system by aldehyde perfusion II. Effect of osmolarity pH of perfusata and fixative concentration. *J. Ultrastruct. Res.* **12,** 187–206.

SCHWAB, D. W., JANNEY, A. H., SCALA, J. & LEWIN, L. M. (1970). Preservation of fine structures in yeast by fixation in a dimethyl sulfoxide-acrolein-glutaraldehyde solution. *Stain Technol.* **45,** 143–7.

SEKI, M. (1936). Zur physikalischen Chemie der Histologischen Färbung IX. Über den Einfluss der Fixierung auf die Färbarkheit der histologische Element. *Z. Zellforsch* **18,** 21–55.

SHEA, S. M. (1971). Lanthanum staining of the surface coat of cells. *J. Cell Biol.* **51,** 611–20.

SHNITKA, T. K. & SELIGMAN, A. M. (1971). Ultrastructural localization of enzymes. *Ann. Rev. Biochem.* **40,** 375–96.

SIPES, J. (1970). El uso del glutaraldehido en escala industrial. *Rev. Assoc. Argent. Quim. Tech. Indust. Cuer.* **11,** 148–55.

SJÖSTRAND, F. S. & BARAJAS, L. (1968). Effect of modifications in conformation of protein molecules on structure of mitochondrial membranes. *J. Ultrastruct. Res.* **25,** 121–55.

THIESSEN, G., THIESSEN, H., DOWIDAT, H. J., LUCIANO, L. & REALE, E. (1970). Die Diffusion der ^{99}Fe-markierten Hämoglobins, ein Artefakt der Glutaraldehyde-Fixierung. *Histochemie* **23,** 171–5.

TOMIMATSU, Y., JANSEN, E. F., GAFFIELD, W. & OLSEN, A. C. (1971). Physical chemical observations on the α-chymotripsin-glutaraldehyde system during formation of an insoluble derivation. *J. Colloid Interface Science* **36,** 51–64.

TOOZE, J. (1964). Measurements of some cellular changes during the fixation of amphibian erythrocytes with osmium tetroxide solutions. *J. Cell Biol.* **22,** 551–63.

TRELSTAD, R. L. (1969). Effect of pH on the stability of purified glutaraldehyde. *J. Histochem. Cytochem.* **17,** 756–7.

TRUMP, B. F. & BULGER, R. E. (1966). New ultrastructural characteristics of cells fixed in a glutaraldehyde-osmium tetroxide mixture. *Lab. Invest.* **15,** 368–79.

TRUMP, B. F. & ERICSSON, J. L. E. (1965). The effect of the fixative solution on the ultrastructure of cells and tissues. A comparative analysis with particular attention to the proximal convoluted tubule of the rat kidney. *Lab. Invest.* **14,** 1245–325.

VANHA-PERTTULA, T. & GRIMLEY, P. M. (1970). Loss of proteins and other macromolecules during preparation of cell cultures for high resolution autoradiography. *J. Histochem. Cytochem.* **18,** 565–75.

VAN HARREVELD, A. & STEINER, J. (1970). Extracellular space in frozen and ethanol substituted central nervous system. *Anat. Rec.* **166,** 117–30.

VAN HARREVELD, A. & KHATTAB, F. I. (1968). Perfusion fixation with glutaraldehyde and post fixation with osmium tetroxide for electron microscopy. *J. Cell Sci.* **3,** 579–84.

WALSH, K. A., HOUSTON, L. L. & KENNER, R. A. (1970). *Chemical Modification of Bovine Trypsinogen and Trypsin in Structure Function Relationships of Proteolytic Enzymes* (eds. P. Desnuelle, H. Neurath & M. Ottesen). Copenhagen: Munksgaard.

WANG, J. H. C. & TU, J. I. (1969). Modification of glycogen phosphorylase by glutaraldehyde. *Biochem.* **8,** 4403–10.

WEBSTER, H. DE F. & AMES, A. (1969). Glutaraldehyde fixation of central nervous tissue: an electron microscopic evaluation in the isolated rabbit retina. *Tissue Cell* **1,** 53–62.

WEBSTER, H. DE F. & AMES, A. & NESBETT, F. B. (1969). A quantitative morphological study of osmotically induced swelling and shrinkage in nervous tissue. *Tissue Cell* **1,** 201–6.

WEST, J. & MANGAN, J. L. (1970). Effects of glutaraldehyde on the protein loss and photochemical properties of kale chloroplasts: preliminary studies on food conversion. *Nature* (*Lond.*) **228,** 466–8.

WRIGGLESWORTH, J. M., PACKER, L. & BRANTON, D. (1970). Organization of mitochondrial structure as revealed by freeze etching. *Biochem. Biophys. Acta* **205**, 125–35.

YOKOTA, K., SUZUKI, Y. & ISHII, Y. (1965). Temperature dependence of the polymerization modes of glutaraldehyde. In: *Chem. Abstr.* **65**, 13835.

YOUNG, E. G. (1963). *Comprehensive Biochemistry*. Vol. 7, p. 25. New York: Elsevier.

Osmolarity of osmium tetroxide and glutaraldehyde fixatives

Q. BONE *and* K. P. RYAN

Marine Biological Association of the UK,
The Laboratory, Citadel Hill, Plymouth

Synopsis. The evidence available to date for the importance of fixative osmolarity is considered together with some observations on the volume changes of crab axons after fixation by osmium tetroxide and glutaraldehyde. The results obtained are compared with those obtained from crab axons and from amphioxus skin cells which had been processed and examined with the electron microscope after initial fixation in fixatives of different composition. It is concluded that the osmolarity of the fixative vehicle is of considerable importance when the fixing agent is glutaraldehyde but is of less importance when the fixing agent is osmium tetroxide or a mixture of the two agents.

Preliminary observations upon crab axons fixed with glutaraldehyde in a vehicle approximating to the internal composition of the cells suggest that this approach to the design of fixative vehicles may be useful.

Introduction

Living cells contain a wide variety of solutes, ranging from small ion species to large molecules, and as they are surrounded by semi-permeable membranes, the cell as a whole is subject to osmotic stress when the concentration of the external medium is changed. Volume changes, therefore, result, owing to the movement of water across the cell membrane, and under certain conditions, the volume of at least some types of cell is directly related to the osmotic pressure of the external solution (see, e.g. Lucké, 1940).

It is the aim of the present survey to consider whether such osmotic effects readily observable with many living cells are of importance during and after the process of fixation when osmium tetroxide and glutaraldehyde are employed as the fixing agents. Evidently, if osmotic effects are not observed with fixed cells it can safely be inferred that the semi-permeable character of the living cell membrane has been destroyed after fixation, whereas if fixed cells *are* subject to osmotic stress, then the fixed cell membrane must be semi-permeable, though it is perhaps improbable that the degree of permeability will remain the same as that for the living cell.

Osmotic forces are large, so that even quite small changes in fixative osmolarity would be expected to produce readily observable volume changes. It would, therefore, be expected that observations at the light microscope level would long ago have provided enough evidence to settle the question of the permeability of the fixed cell membrane.

Hertwig (1931) observed that eggs of the sea urchin *Arbacia* increased markedly in volume when placed in a 4% solution of formaldehyde in distilled water, but nearly retained their original volume if placed in a 4% formaldehyde solution made up in sea-water; in accord with these direct observations, Young (1935) found that neurons of *Sepia* were much better fixed by 'seawater-formalin' (as judged by the appearance of stained sections) than by formalin in distilled water. He also observed that better fixation was obtained with an isosmotic vehicle when the fixing agent was picric acid, chromic acid, or, interestingly enough, osmium tetroxide. Young concluded that it was necessary to use a fixative vehicle isosmotic with the normal external medium, neglecting the contribution to the osmolarity of the fixative made by the fixing agent itself. Both Hertwig and Young had examined marine material, where the difference between fixatives made up in distilled water and in vehicles isosmotic with the medium are greatest. But in a similar way, Baker (1958) (who gives a valuable discussion of the opinions of light microscopists on fixative osmolarity, and of the value of so-called 'indifferent' salts in fixation) found fixation was better when indifferent salts were added to formaldehyde and osmium tetroxide fixatives for mammalian material. The best results seemed to have been obtained when the concentration of these salts was such as to make the fixative vehicle of slightly lower osmolarity than the body fluids. For example, addition of sodium chloride to the fixative to a final concentration of 0.7% gave better results when formaldehyde was used as the fixing agent than if the concentration of sodium chloride was increased to make the vehicle isosmotic with the body fluids. As he had long pointed out (Baker, 1933), many of the fixatives used for light microscope work have a total osmolarity much above that of the body fluids.

The evidence available from light microscope observations thus suggested that cells fixed both with formaldehyde and with osmium tetroxide were subject to osmotic stress, and that the best fixation was obtained when the fixing agents were made up in vehicles of similar osmolarity to the normal fluids external to the tissues. Certainly, some workers observed anomalies to this general rule, as Crawford & Barer (1951) noted in their interesting study of formaldehyde fixation, but in general most workers add indifferent salts to fixatives for light microscope studies, particularly for marine material.

It is remarkable that at the more critical electron microscope level, the evidence available at present is conflicting, and that the vehicles in common use for osmium tetroxide and glutaraldehyde fixatives differ a good deal in osmolarity even when employed on similar material. For example, some workers consider that the total osmolarity of glutaraldehyde fixatives should be equivalent to that of the body fluids, that is, the contribution of the fixing agent itself to the total osmolarity of the fixative is taken into account. Other workers ignore this contribution, and like many light microscopists, make up their fixative vehicles to be isosmotic with the body fluids. Again, some workers have obtained good fixation using osmium tetroxide in distilled water, whereas others use this fixing agent in physiological saline solutions. Some of the fixatives that have been used recently on different materials are compared in Table 1 from which it is clear that there is a diverse range of opinion, especially in connection with osmium tetroxide. The dif-

Table 1. Some fixatives containing glutaraldehyde and osmium tetroxide employed in electron microscope investigations

Authors	*Material*	*Fixative vehicle*	*Osmolarity vehicle*	*Total osmolarity*	*Remarks*
	Glutaraldehyde & formaldehyde				
Baker (1965)	Mammalian	Sucrose or NaCl	Below body fluids	Above body fluids	
Karnovsky (1965)	Various	Buffer	Below body fluids	Much above	
Maunsbach (1966)	Mammalian	Various	Below body fluids	Above body fluids	Best of many tested
Tilney & Goddard (1970)	*Arbacia* egg	Seawater	As external medium	Above	
Perrachia (1970)	Fresh-water crayfish	Buffer	Below body fluids	Above	
Ovalle (1970)	Mammalian	Buffer	Below body fluids	As body fluids	
Busson-Mabillot (1971)	Fresh-water teleost	Buffer	Below body fluids	As body fluids	Best of many tested
	Osmium tetroxide				
Sjöstrand (1956)	Mammalian	Physiological saline	As body fluid		
Malhotra (1962)	Mammalian	Distilled water	Much below body fluids		Osmolarity unimportant
Baker (1965)	Mammalian	Distilled water	Much below body fluids		Osmolarity unimportant
Chuang (1968)	Marine invertebrates	Seawater or distilled water	As body fluids or much below		Identical results
Busson-Mabillot (1971)	Fresh-water teleost	Distilled water	Much below body fluids		Reasonable results
Busson-Mabillot (1971)	Fresh-water teleost	Buffer + sucrose	Slightly above body fluids		Best results

ferent osmolarities of the fixatives and fixative vehicles employed presumably reflect different opinions of the state of the fixed cell membrane. Some workers have investigated this state directly, either by electrical methods, or by observing the volume changes of fixed cells in different solutions. Elbers (1966), for example, placed eggs of the freshwater snail *Limnaea* in conductivity cells and measured the rate of loss of ions across the cell membrane as fixing agents were applied. He found that after glutaraldehyde fixation, ions were lost very slowly from fixed cells, but that cells fixed in osmium tetroxide very quickly became freely permeable to ions (within seconds). He concluded from such observations that 'fixation by glutaraldehyde does not appreciably change the permeability to ions of the cell membrane of the *Limnaea* egg'; in contrast, osmium tetroxide fixation renders the cell membrane freely permeable to salts so that, as he pointed out, 'it is not clear what part the frequently used tonicity additives in osmium tetroxide fixatives could play in the fixation process of single cells'.

Millonig (1968) repeated Hertwig's experiments on sea urchin eggs, using glutaraldehyde as the fixing agent, and found them to be osmotically sensitive after fixation. The original volume of the eggs was best maintained by fixation in a solution containing 2% sodium chloride, so that the fixative vehicle had a lower osmolarity than sea water.

Denton and one of us (Bone & Denton, 1971) used volume changes of the reflecting cells on the scales of marine teleosts (manifested by changes in the wavelength of light reflected by the array of guanine platelets within the cells) to show that these cells were osmotically active after glutaraldehyde fixation. The fixed cells appeared to be impermeable to sodium and to potassium, as they were in life, and it was found that after fixation, the cells retained their original volume in life when placed in solutions of lower osmolarity than the body fluids. Furthermore, the original cell volume could best be maintained by using a fixative in which the osmolarity of the fixative vehicle was some two-thirds of that of the body fluids. Thus, although the total osmolarity of the fixative was above that of the body fluids, the contribution to fixative osmolarity made by the fixing agent could be ignored. It was not possible to demonstrate with this material that there were no volume changes after osmium tetroxide fixation or post-treatment, but this seemed probable.

In direct contrast to these observations, Carstensen *et al.* (1971) concluded from dielectric measurements upon mammalian red cells that glutaraldehyde-fixed cells are not subject to osmotic stress, the cell membranes ceasing to be osmotic barriers immediately after fixing. These workers were, in fact, able to vary the external ionic concentration by a factor of ten without altering the volume of the fixed cells, although Ponder (1940) had previously found that formalin-fixed red cells were osmotically sensitive.

It is difficult to imagine that such conflicting results can simply be explained by different effects of the same fixing agent upon different types of cell membranes; it certainly seems more probable that the loss of sensitivity to osmotic stress after fixation found by Carstensen *et al.* resulted from the damage to the membranes of the red cells they were investigating.

These few direct investigations of the membrane permeability of fixed cells suggest that it should not be necessary to consider osmotic effects with fixatives containing osmium tetroxide, although osmotic effects are likely to occur with fixatives containing glutaraldehyde, when fixative osmolarity will be important, as well as the osmolarity of subsequent solutions. As will be discussed in a later section, this is too simple a view, and it seems that osmolarity has to be taken into account even with fixatives containing osmium tetroxide.

What follows is a report of results obtained by direct observation of volume changes in fixed cells, and indirect observations on cells examined with the electron microscope after fixation in fixing mixtures of different osmolarities.

Materials and methods

Leg nerves of the shore crab *Carcinus maenas* (L.) were dissected out and 1–2 cm lengths placed in seawater in a small chamber on a microscope slide. One or more of the largest axons, which are up to 50 μm in diameter, were separated from the nerve bundles, and their diameter measured in different solutions using a micrometer eyepiece. Volume changes produced by the osmotic effects of different solutions change the length of the fibres to some extent, but for the present purpose it was assumed that the volume of the fibre was related directly to its diameter. It was found possible to maintain such fibres for several hr in a suitable condition for volume measurements. *Carcinus* axons survive well in seawater, which is approximately isosmotic with the blood; the osmolarity of the different seawater solutions in which they were placed was altered by dilution with distilled water, or by the addition of sodium chloride.

Table 2. Composition of fixatives employed

Fixing agent	*Vehicle*	*Osmotic Pressure of vehicle (mosM)*
5% glutaraldehyde	Crab-Ringer + distilled water or sucrose	500; 750; 1080; 1250; 1500
	0.2 M *s*-collidine buffer + sucrose	500; 750; 1080; 1250; 1500
	400 mM KCl + sucrose + 5% polyethylene glycol (mol. wt. 10 000)	1080 (+polyethylene glycol)
1% Osmium tetroxide	Distilled water	
	Crab-Ringer	1080
	Crab-Ringer + sucrose	1500
	400 mM KCl + sucrose + 5% polyethylene glycol (mol. wt. 10 000)	1080 (+polyethylene glycol)
Simultaneous 2.5% glutaraldehyde + 1% osmium tetroxide	Crab-Ringer + distilled water or sucrose	500; 1080; 1500

Other leg nerves were fixed in 5% glutaraldehyde dissolved in seawater, or in 1% osmium tetroxide in seawater, for various periods, before large axons were dissected free and similar experiments performed.

Leg nerves of the spider crab *Maia squinado* (Herbst) were used to study the effects, as judged by examination of electron micrographs, of varying fixative osmolarity. Most of the axons in *Maia* leg nerves are of similar diameter (Abbott *et al.*, 1958), so that it is simpler to find comparable fields on different grids than with the wider fibre spectrum of the *Carcinus* nerves. The nerves were fixed attached to small pieces of card, detached and rinsed in the vehicle before being dehydrated and embedded in Taab resin. Those fixed in glutaraldehyde alone were post-osmicated in 1% osmium tetroxide in distilled water. Sections stained with lead citrate (Reynolds, 1963) were examined in either a Philips EM 200 or an EM 300. The compositions of the different fixatives employed are given in Table 2. Secondly, small portions of the mid-lateral region of the body of amphioxus (*Branchiostoma lanceolatum* Pallas) were fixed in solutions of different osmolarity and processed in the same way as the crab fibres. The effects on the skin cells were examined in this material. The different fixatives employed are given in Table 2.

Results

Volume changes of Carcinus axons

Living *Carcinus* axons change in volume, as one would expect, when placed in solutions of different osmolarity, whether these solutions contain salts or sucrose. The axons do not behave as perfect osmometers, partly because simple measurement of diameter does not take into account the length changes which certainly occur (see Hill, 1950), and partly because the axoplasm tends to escape from the cut ends of the axons as they swell in dilute solutions. Fig. 1 illustrates the results obtained when axons are placed in

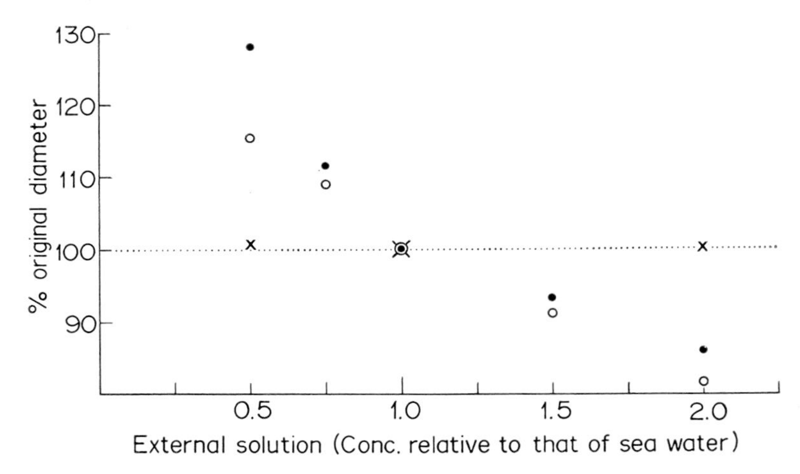

Figure 1. Percentage of original diameter of *Carcinus* axons in different seawater solutions. Open circles (○): living axon; closed circles (●): glutaraldehyde-fixed axon; crosses (×): osmium-tetroxide-fixed axon.

solutions more concentrated and more dilute than seawater. The open circles represent the percentage of the original diameter of a living axon in different concentrations of seawater. The changes are reversible, and take place within a few minutes of changing the external solution. The closed circles show the percentage of the original diameter of an axon which had been fixed in 5% glutaraldehyde in seawater for 50 min before changing the external solutions. These changes were also reversible. It is plain that the fixed axon was subject to osmotic stress. Axons fixed for 48 hr in 5% glutaraldehyde in seawater were still subject to osmotic stress, swelling and shrinking when placed in sucrose solutions of different osmolarity to seawater. In contrast to these results with glutaraldehyde fixed axons, the crosses show the percentage of the original diameter of an axon that had been fixed in 1% osmium tetroxide in seawater for 90 min before changes of the external solution. Little diameter change took place when the external solutions were changed (although the axon swelled slightly during the period in the fixing solution), and similar results were obtained when osmium tetroxide-fixed axons were placed in sucrose solutions of differing osmolarity. It, therefore, appears that after osmium tetroxide fixation, the axon membrane is freely permeable not only to smaller ions, but also to sucrose molecules.

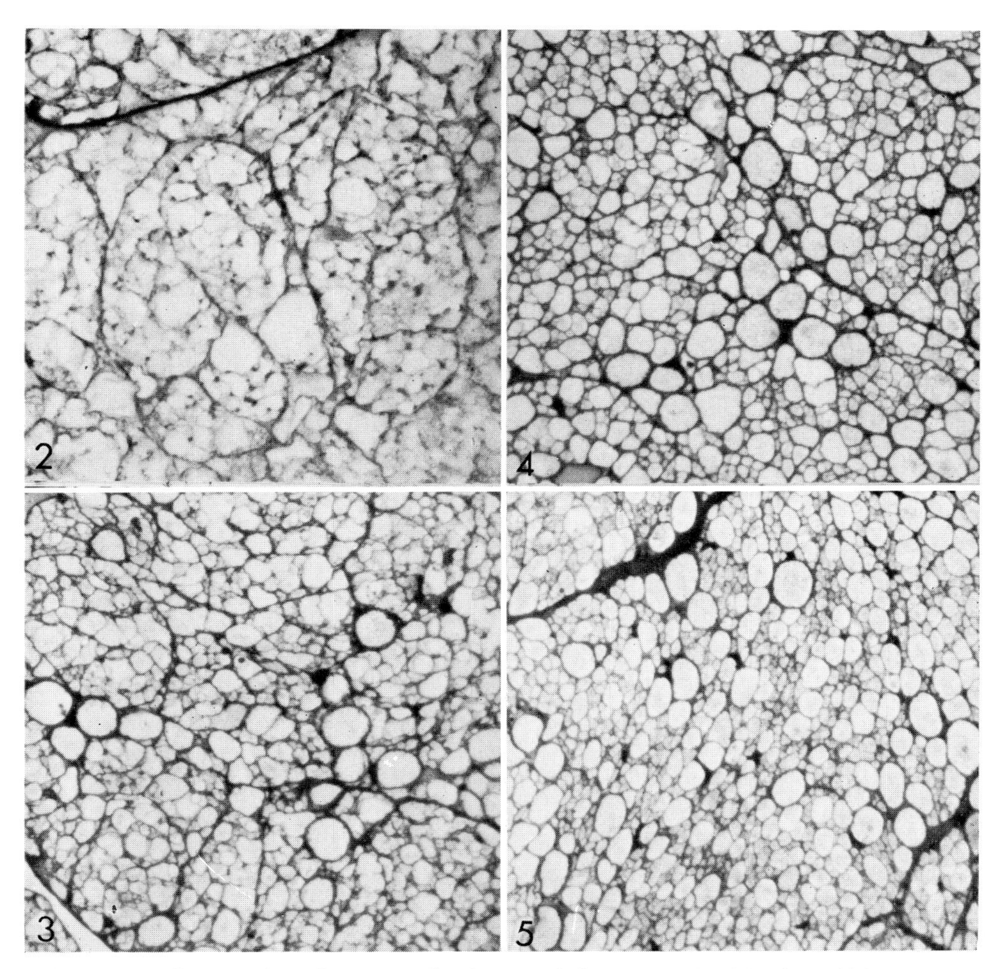

Figures 2–5. Cross-section of axon bundles from *Maia* leg nerves, fixed in glutaraldehyde, embedded in Taab resin, cut at 1 μm, and stained with Toluidine Blue. All × 400. Osmolarity of fixative vehicles:

Fig. 2: 500 mosM; *Fig.* 3: 1080 mosM; *Fig.* 4: 1250 mosM; and *Fig.* 5: 1500 mosM.

In these experiments, it was not possible to examine transient volume changes, lasting up to 10–20 sec, and as will be seen in a later section, these may be of importance in fixation. However, so far as they go, our direct observations on membrane permeability after treatment with either fixing agent are in agreement with previous results on *Limnaea* eggs and teleost reflecting cells.

Glutaraldehyde fixation

As expected from the observations on *Carcinus* axons, the osmolarity of the fixative vehicle produces obvious effects on *Maia* axons post-osmicated and processed for electron microscopy. Even at the light microscope level (Figs. 2–5) low osmolarity fixative vehicles may be seen to yield swollen axons, whereas high osmolarity vehicles yield somewhat shrunken axons with wider interspaces. When examined with the electron microscope (Figs. 7–11) these effects are more conspicuous. Axons fixed in low osmolarity vehicles are swollen and have a typically 'empty' appearance; there is little space between axons. Those axons fixed in vehicles of high osmolarity are somewhat crenated, and inter-axonal spaces are larger. The appearance of axons fixed in unbuffered 1% osmium tetroxide dissolved in seawater is shown in Fig. 6. This fixative gives relatively poor results for crab axons; they all appear much shrunken and crenated.

Similar evidence for the importance of osmolarity in glutaraldehyde fixation is provided by the effects of different fixatives upon amphioxus skin cells (Figs. 12–20). These cells are cuboidal, and interdigitate to form a single layer of cells over a connective tissue lattice. At their bases are small elongate vesicles (which Olsson, 1961, has shown to contain glycogen), and at the free border, in contact with seawater, are larger mucoid vesicles under an array of microvilli. When fixed in vehicles of higher osmolarity than seawater (the animal is in life probably isosmotic with seawater), the skin cells shrink, discharge the contents of their basal vesicles and shrink away from the basal membrane. The surface microvilli become thin and elongate. In fixative vehicles of lower osmolarity than seawater, the skin cells apparently swell, so that the microvilli are now short and broad, and the base of the cells remain in contact with the basal membrane with undischarged basal vesicles.

Taken together with the direct observations on volume changes of fixed cells, our observations on processed material of *Maia* and amphioxus provide a reasonably clear further indication that fixative osmolarity is of importance when the fixing agent is glutaraldehyde. All but the fixatives made up in 500 mosM vehicles had a total osmolarity above that of the external medium, and it appears that the contribution to total osmolarity made by the fixing agent can be neglected, as Young (1935) suggested.

Osmium tetroxide fixation

Maia axons appear to be freely permeable to salts and to sucrose after osmium tetroxide fixation, and it seemed probable that the osmolarity of the fixative vehicle should be of little importance in osmium tetroxide fixatives. Yet most workers post-osmicate glutaraldehyde-fixed material in osmium tetroxide solutions to which salts or buffers are added; they state, as in the case of Olsson's (1961) investigations of amphioxus skin cells, that rinsing osmium-fixed tissue in distilled water gives very poor results. Our own observations on amphioxus skin cells fixed in osmium tetroxide solutions of different osmolarity did not indicate that the osmolarity of the vehicle is of importance; similar

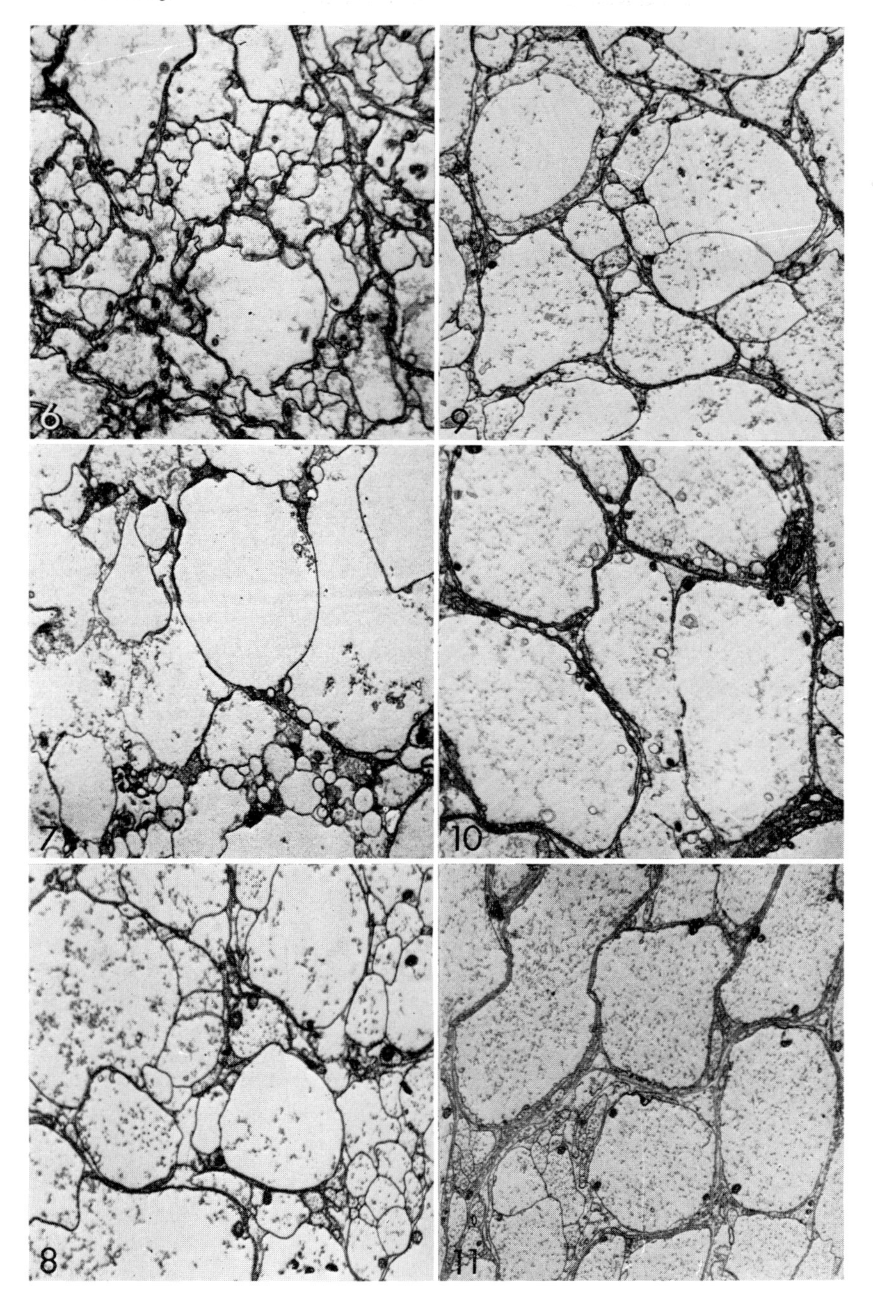

Figure 6. *Maia* axons fixed in 1% osmium tetroxide in seawater. × 3400
Figures 7–11. *Maia* axons fixed in glutaraldehyde at different osmolarity. All × 3400
Osmolarity of vehicle: *Fig.* 7: 500 mosM; *Fig.* 8: 750 mosM; *Fig.* 9: 1080 mosM; *Fig.* 10: 1250 mosM; and *Fig.* 11: 1500 mosM.

results were obtained from material fixed in osmium tetroxide in distilled water, and in seawater with added sucrose. *Maia* axons (Figs. 21, 22) fixed in distilled water as a vehicle often show vesicles in the axoplasm that are usually not found in axons fixed with isosmotic or hyperosmostic vehicles. They also show more clearly vesiculation of the axon membrane, a characteristic artefact of osmium tetroxide fixation seen in mammalian material (Tormey, 1964; Maunsbach, 1966), and in crustacean nerve sheaths (Doggenweiler & Heuser, 1957). However, in both axons fixed in high and low osmolarity vehicles, similar appearances are observable in different fields, and as compared with glutaraldehyde-fixed material, the axons appear more 'empty', since microtubules are not clearly seen after osmium tetroxide fixation at any osmolarity.

Our observations on osmium tetroxide-fixed material do not indicate that the osmolarity of the fixative vehicle is of consequence with this fixing agent.

Glutaraldehyde and osmium tetroxide

When mixtures of glutaraldehyde and osmium tetroxide are used for the fixation of *Maia* axons, the results (Figs. 23, 24) are rather similar to those obtained with osmium tetroxide alone, in that the osmolarity of the fixative vehicle does not seem to be of importance; similar results are obtained with vehicles of high and low osmolarity. It might be supposed that glutaraldehyde would penetrate more rapidly than osmium tetroxide (in view of Hopwood's (1970) observations on tissue blocks), but in *Maia* leg nerves, where the axons are loosely held in small bundles, this does not appear to be the case. At both high and low osmolarity fixation, the same membrane vesiculation artefact is found that is seen when axons are fixed by osmium tetroxide alone. Since this artefact is not observed when axons are fixed in glutaraldehyde and then post-osmicated, it seems that in these small cell groups at least, both fixing agents act upon the cell membrane together.

Microtubules within the axoplasm are perhaps better preserved by the two fixing agents acting together than in axons fixed by osmium tetroxide alone, but it does not seem with our material that there is any significant advantage to be gained by adding osmium tetroxide to glutaraldehyde fixatives.

'Internal' fixative vehicles

Some preliminary observations have been made using fixatives made up so that the vehicle is isosmotic with the external fluid around the cell, but is in approximate *ionic* equilibrium with the *inside* of the cell. The internal constituents of *Maia* axons are not known, but by analogy with those of *Homarus* (Brinley, 1965) are presumably rich in potassium and low in sodium and chloride, and also contain large organic anions. The fixative vehicle used was based upon the internal constituents of *Sepia* axons, but

---

Figure 12. Amphioxus skin cell fixed in seawater-glutaraldehyde to show general features of cell. × 9500

Figures 13–15. Microvilli of free border of skin cell fixed with glutaraldehyde. All × 17 200
Osmolarity of vehicle: *Fig.* 13: 500 mosM; *Fig.* 14: 1080 mosM; and *Fig.* 15: 1500 mosM.

Figures 16–20. Basal region of skin cells fixed in glutaraldehyde. All × 17 200
Osmolarity of vehicle: *Fig.* 16: 500 mosM; *Fig.* 17: 750 mosM; *Fig.* 18: 1080 mosM; *Fig.* 19: 1250 mosM; and *Fig.* 20: 1500 mosM.

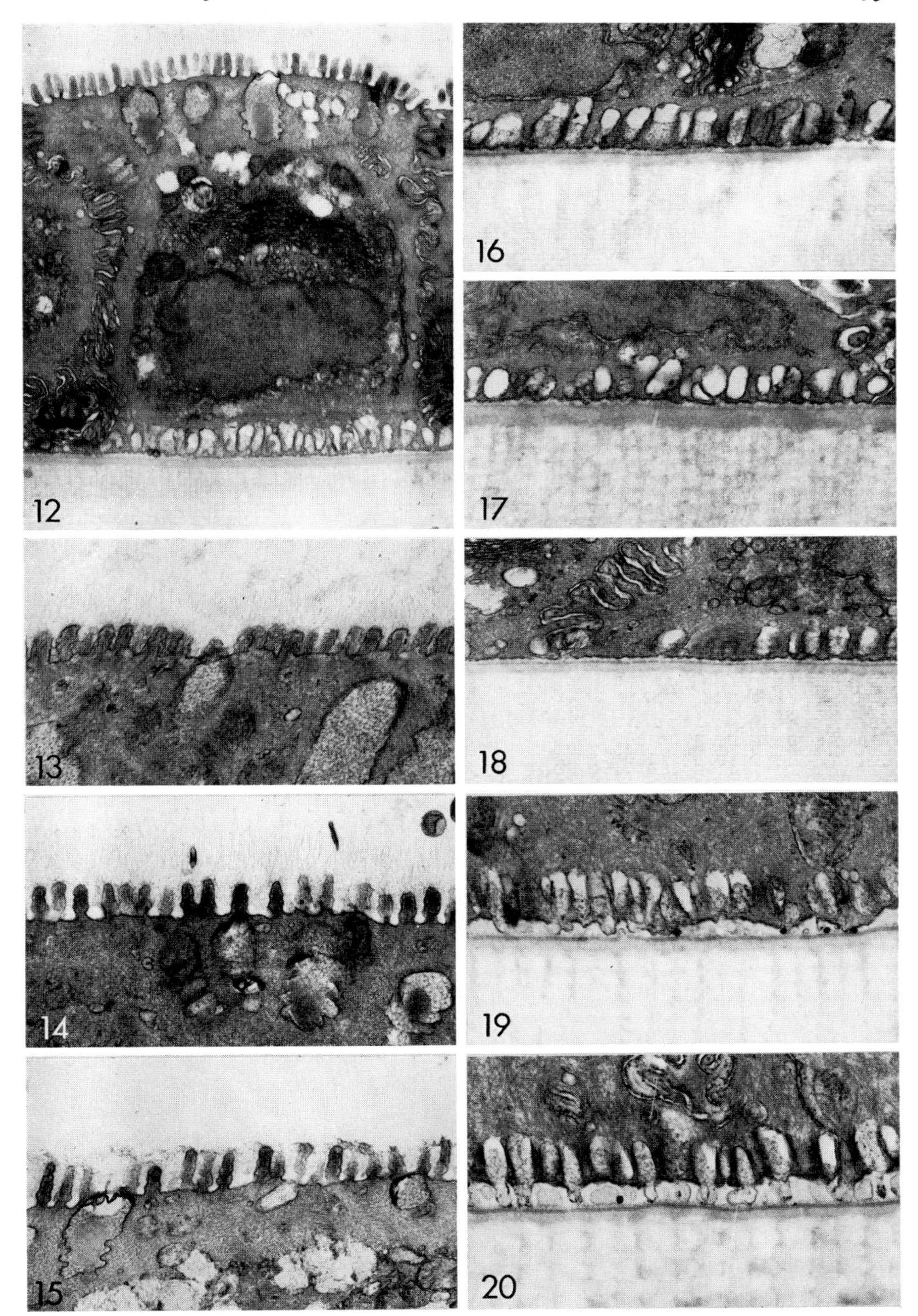
12
13
14
15
16
17
18
19
20

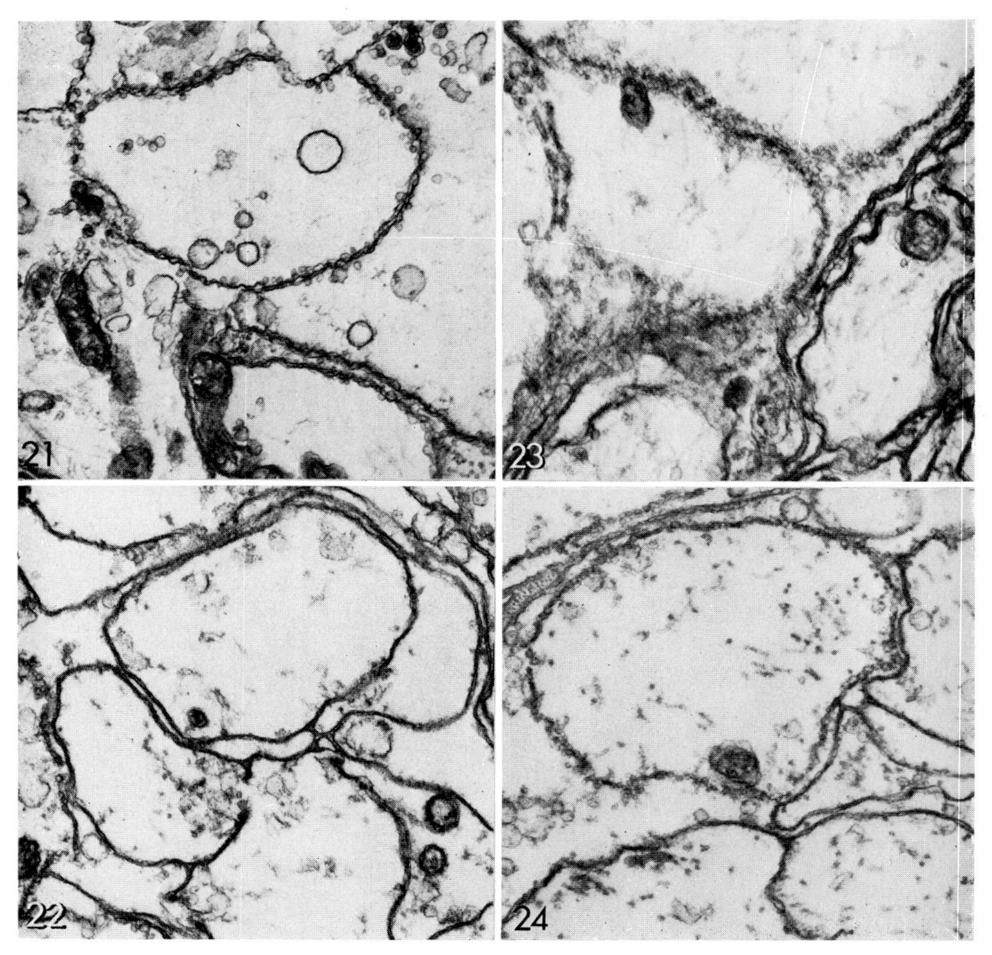

Figure 21. *Maia* axons fixed in distilled water-osmium tetroxide. × 19 000
Figure 22. *Maia* axons fixed in seawater-osmium tetroxide. × 19 000
Figure 23. *Maia* axons fixed in glutaraldehyde-osmium tetroxide, osmolarity of vehicle: 500 mOSM. × 19 000
Figure 24. *Maia* axons fixed in glutaraldehyde-osmium tetroxide, osmolarity of vehicle: 1500 mOSM. × 19 000

contained a high chloride content, some sucrose to adjust the osmolarity, and 5% polyethylene glycol (Table 2). The latter shrinks osmium tetroxide-fixed axons, and thus is apparently unable to cross osmium tetroxide-fixed membranes.

Axons fixed in this vehicle using osmium tetroxide as the fixing agent (Fig. 26) are not perfectly preserved, but there is less shrinkage than with a seawater vehicle (Fig.25), and the smaller axons in the field appear less 'washed out'. The vesiculation artefact may still sometimes be observed, but it is less common than when axons are fixed in seawater or distilled water as the vehicle. When the fixing agent is glutaraldehyde (Fig. 28) fixation is similar to that obtained by isosmotic glutaraldehyde fixation using a seawater vehicle

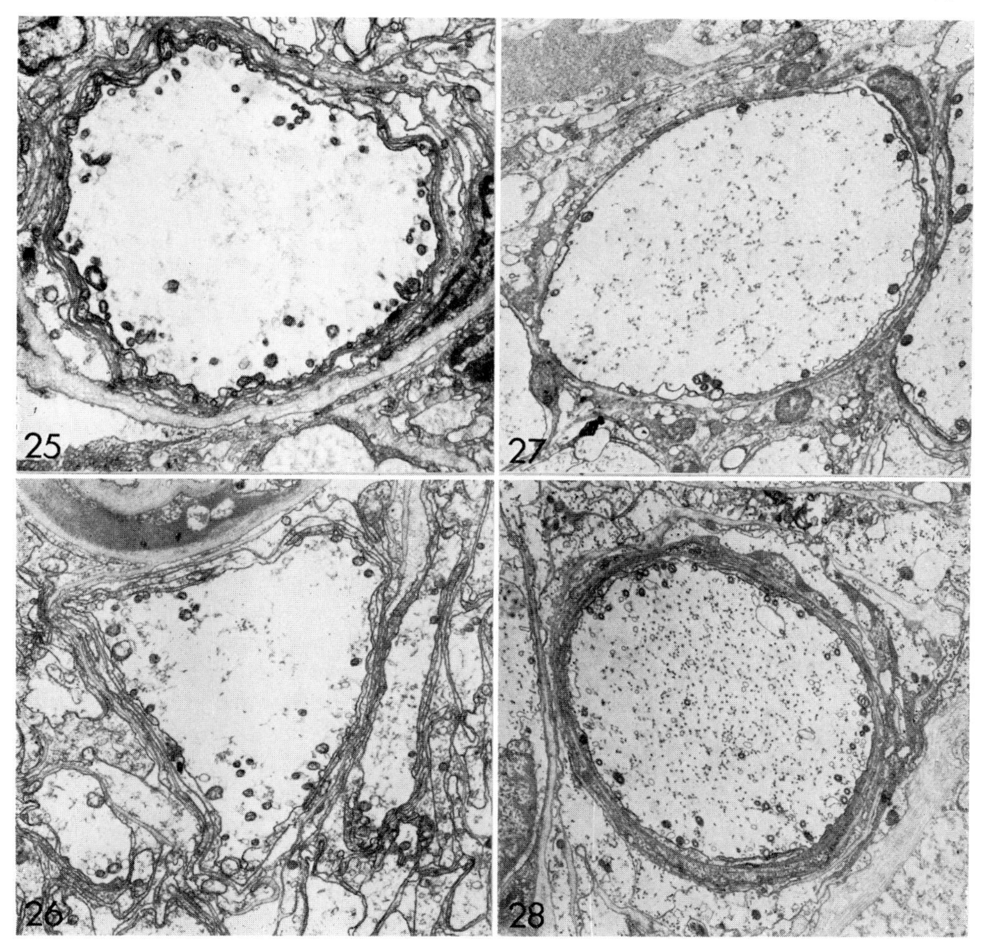

Figure 25. Large axon from *Maia* leg nerve fixed in seawater-osmium tetroxide. × 4700
Figure 26. Large axon from *Maia* leg nerve fixed in 'internal' vehicle osmium tetroxide. × 4700
Figure 27. Large axon from *Maia* leg nerve fixed in seawater-glutaraldehyde. × 4700
Figure 28. Large axon from *Maia* leg nerve fixed in 'internal' vehicle glutaraldehyde. × 4700

(Fig. 27) and microtubules are well preserved, although there are now small vesicles in the axoplasm. In the smaller axons, microtubules are more numerous in material fixed in the 'internal' vehicle.

Discussion

Living cells do not possess ideal semi-permeable membranes, since their membranes normally permit passage of some solutes as well as water; it is for this reason that isosmotic external solutions of different ions may not be isotonic for a particular cell type. Some solutes, such as urea or glucose, are usually very freely permeable and thus if added to the normal external medium would be expected to cause only small transient volume changes before the solute is equally distributed across the cell membrane. Less freely penetrating solutes will produce larger initial volume changes, later reversed as equilibrium is finally attained. Since there are very few measurements of the relative permeabilities of fixed cell membranes to different solutes, it is at present impossible to design fixatives on a rational osmotic basis when the fixative vehicle contains diverse ion species (e.g. indifferent salts and buffers), as it is probable that the process of fixation will change the effective pore size of the cell membrane.

It is known that for fixed cell membranes, effects are observed which would be expected if the cell membrane is not ideally semi-permeable. Thus, transient changes have been observed with glutaraldehyde-fixed teleost reflecting cells when urea is added to the external medium.

However, despite the lack of direct observations upon the permeability of fixed cell membranes to different solutes, it is possible to draw some conclusions from the empirical data provided by observations upon processed material. Most studies on different material using glutaraldehyde as the fixing agent indicate that the volume changes attributable to fixation are least if the fixative vehicle is slightly below the osmolarity of the normal external medium, and if the osmotic pressure of the fixing agent is neglected. The extended series of careful observations by Maunsbach (1966) and Busson-Mabillot (1971) on the effects of varying fixative osmolarity lead to this conclusion; they are supported by other workers who use aldehyde as the fixing agent and design their fixative to have a total osmolarity equivalent to that of the body fluids. Several explanations may be suggested for this finding; some have been considered by Bone & Denton (1971). Perhaps the most probable is that large anions within the cell may be condensed under the cross-linking action of the fixing agent, thus reducing the osmotically active constituents within the cell.

Cells fixed with osmium tetroxide are apparently freely permeable to small solutes, so that it is understandable that several workers have obtained good fixation using osmium tetroxide in distilled water. However, it is likely that in most cells, and certainly in nerve, muscle, and red cells, there will be larger solutes to which even the osmium tetroxide-fixed cell membrane is impermeable (as it is to polyethylene glycol). On simple osmotic grounds, therefore, cells fixed in osmium tetroxide in distilled water would be expected first to rapidly lose their small solutes to the surrounding medium, and then to gradually swell as a result of the unequal distribution of the internal impermeable large solutes across the membrane. We have observed swelling of axons during a period of an hour in osmium tetroxide fixatives, and Bahr *et al.* (1957) noted that tissue slices immersed in osmium tetroxide of various fixative vehicle osmolarity all showed marked swelling. An obvious way of avoiding osmotic swelling resulting from large internal solutes is to add a large solute to the fixative vehicle, and this has been done by several workers. When using glutaraldehyde as the fixing agent, Bohman & Maunsbach (1970) added dextran or polyvinylpyrrolidone (PVP) at 2% or 4% to the fixative vehicle and to the osmium tetroxide solution used for post-osmication.

Robertson & Schultz (1970) employed 1% gum acacia similarly. Since glutaraldehyde-fixed membranes seem to be impermeable to sucrose, the direct osmotic effects upon cells of adding such substances (in preference to salts or sucrose) to glutaraldehyde fixatives are unlikely to be of consequence until the post-osmication stage. Moreover, at the concentrations employed of these high mol. wt. substances, their contribution to fixative osmolarity is very small. However, Robertson & Schultz were fixing by perfusion, and as Bohman & Maunsbach suggested, the effects observed may be due to changes in osmotic balance across the capillary wall. Del Cerro & Snider (1972), working with similar material, obtained inconclusive results when they added PVP to aldehyde fixatives, and found the tonicity of the buffers employed to be the most significant variable among many studied. Our own preliminary observations upon *Maia* axons suggest that volume change is less when a large impermeable solute is added to osmium textroxide fixatives and the cells are fixed by immersion.

Recent work on the dimensions of the transverse tubular system in frog muscle fibres, treated with sucrose-Ringer solutions and fixed in osmium tetroxide and acrolein fixatives (Birks & Davey, 1972), has demonstrated that this system (in the living cell in connection with extra-cellular space) is apparently osmotically isolated from extra-cellular space when fixed by osmium tetroxide, but not when fixed by acrolein or glutaraldehyde. Birks & Davey suggest that their results (apparently arguing for the retention of semi-permeability by osmium-tetroxide-fixed membranes) are to be explained in terms of the isolation of the T-system from extra-cellular space during osmium tetroxide fixation owing to disruption artefacts closing the connections of the T-tubules with the outside of the fibres. It seems probable that if this explanation is correct, the effect must be a transient one, before the permeability increase resulting from osmium tetroxide fixation of the T-tubules themselves.

Little is known of the transient osmotic effects that will result from equilibration of reasonably freely permeable solutes across fixed cell membranes. It is for this reason that it may be worthwhile to investigate further the possibility of fixing cells in vehicles which are similar ionically to their internal constituents. Where there is in life a high concentration gradient of certain ions across the cell membrane, these ions are likely to possess different activity co-efficients, and so the process of equilibration that must take place after fixation may produce osmotic stresses and volume changes. Thus, Hill (1950) has shown that in *Sepia* axons exchange across the membrane of sodium for internal potassium will produce osmotic swelling, and any cell that possesses a low sodium content in life, because it actively pumps out sodium, may be better fixed by vehicles containing little sodium, rather than by using physiological saline as the vehicle. The duration of fixation will be of importance here, as it appears that the equilibration of ions across glutaraldehyde-fixed membranes must be a slow process, and that glutaraldehyde-fixed cells are still osmotically sensitive after long periods in the fixative.

Our experience with 'internal' fixative vehicles is limited, but our preliminary results suggest that further work on 'internal' vehicles could yield useful results, as Crawford & Barer (1951) had concluded from their studies of formaldehyde-fixed cells. It is interesting that Elford & Walter (1972) have found low sodium-high potassium media to be the most successful for incubation of mammalian muscle at low temperatures, when the normal ionic pumping mechanisms are presumably inactivated, as they must be by fixation.

So far, fixative osmolarity has been considered in terms of the effects of fixatives upon single cells or small groups of cells. It is plain that when tissues or large cell groups are fixed, conditions will not be as simple, and it is perhaps for this reason that there are conflicting opinions about, for example, the osmolarity of osmium tetroxide fixatives. In general, it is probably correct to suppose that for immersion fixation for electron microscopy, the pieces of tissue taken are sufficiently small that indirect effects of fixative osmolarity are unlikely to be of importance. However, it would nevertheless seem prudent to design the vehicles for osmium tetroxide fixatives and for post-osmicating solutions to be isosmotic with the normal extracellular environment, and reasonable to add large impermeable solutes to such solutions. For glutaraldehyde fixation, the vehicle must be close to the osmolarity of the normal external solution, and although no evidence has been obtained during the present study, it seems that the most successful fixation is usually obtained when the osmolarity of the vehicle is somewhat below that of the normal external medium.

Finally, volume changes resulting from fixation are certainly not all due to simple osmotic effects, as Elbers (1966) pointed out in an interesting discussion of the changes taking place as a result of fixation with osmium tetroxide, and as Crawford & Barer (1951) also emphasized. However, since osmotic forces are large, it is reasonable to attempt to minimize their action during fixation procedures. It is unfortunate that in the present state of knowledge it is not possible to design fixatives on theoretically reasonable osmotic grounds, even for isolated single cells (as Elbers has also concluded): more information is needed about the permeability of fixed cell membranes before we can successfully design fixatives to avoid osmotic stress during fixation.

Acknowledgement

We are grateful to Miss Myra Bowkett for photographic assistance.

References

ABBOTT, B. C., HILL, A. V. & HOWARTH, J. V. (1958). The positive and negative heat production associated with a nerve impulse. *Proc. Roy. Soc. Lond. B* **148,** 149–87.

BAHR, G. F., BLOOM, G. & FRIBERG, U. (1957). Volume changes of tissue in physiological fluids during fixation in osmium tetroxide or formaldehyde and during subsequent treatment. *Exp. Cell Res.* **12,** 342–55.

BAKER, J. R. (1933). *Cytological technique.* London: Methuen.

BAKER, J. R. (1958). *Principles of biological microtechnique.* London: Methuen.

BAKER, J. R. (1965). The fine structure produced in cells by fixatives. *J. Roy. micr. Soc.* **84,** 115–31.

BIRKS, R. I. & DAVEY, P. F. (1972). An analysis of volume changes in the T-tubes of frog skeletal muscle exposed to sucrose. *J. Physiol., Lond.* **222,** 95–111.

BRINLEY, F. J. (1965). Sodium, potassium and chloride concentrations and fluxes in the isolated giant axon of *Homarus. J. Neurophysiol.* **28,** 742–72.

BOHMAN, S-O. & MAUNSBACH, A. B. (1970). Effects on tissue fine structure of variations in colloid osmotic pressure of glutaraldehyde fixative. *J. Ultrastruct. Res.* **30,** 195–208.

BONE, Q. & DENTON, E. J. (1971). The osmotic effects of electron microscope fixatives. *J. Cell Biol.* **49,** 571–81

BUSSON-MABILLOT, S. (1971). Influence de la fixation chimique sur les ultrastructures. I. Etude sur les organites du follicule ovarien d'un poisson téléostéen. *J. Microsc.* **12,** 317–48.

CARSTENSEN, E. L., ALDRIDGE, W. G., CHILD, S. Z., SULLIVAN, P. & BROWN, H. H. (1971), Stability of cells fixed with glutaraldehyde and acrolein. *J. Cell Biol.* **50,** 529–32.

CRAWFORD, G. N. C. & BARER, R. (1951). The action of formaldehyde on living cells as studied by phase-contrast microscopy. *Quart. J. micr. Sci.* **92,** 403–52.

CHUANG, S. H. (1968). Seawater and osmium tetroxide fixation of marine animals. In: *Cell structure and its interpretation* (eds. S. M. McGee Russell & K. F. A. Ross). St Martin's Press: New York.

DEL CERRO, M. P. & SNIDER, R. S. (1972). Studies on the developing cerebellum. II. The ultrastructure of the external granular layer. *J. comp. Neurol.* **144**, 131–64.

DOGGENWEILER, C. F. & HEUSER, J. E. (1967). Ultrastructure of the prawn nerve sheaths. Role of fixative and osmotic pressure in vesiculation of thin cytoplasmic laminae. *J. Cell Biol.* **34**, 407–20.

ELBERS, P. F. (1966). Ion permeability of the eggs of *Limnaea stagnalis* L. on fixation for electron microscopy. *Biochim. biophys. Acta* **112**, 318–19.

ELFORD, B. C. & WALTER, C. A. (1972). Preservation of structure and function of smooth muscle. cooled to −79°C in unfrozen aqueous media. *Nature, Lond.* **236**, 58–9.

HERTWIG, G. (1931). Der Einfluss der Fixierung auf das Kern- und Zell-volumen. *Z. Mikr. anat. Forsch* **23**, 484–504.

HILL, D. K. (1950). The volume change resulting from stimulation of a giant nerve fibre. *J. Physiol., Lond.* **111**, 304–27.

HOPWOOD, D. (1970). The reactions between formaldehyde, glutaraldehyde and osmium tetroxide, and their fixation effects on bovine serum albumin and on tissue blocks. *Histochemie* **24**, 50–64.

KARNOVSKY, M. J. (1965). A formaldehyde-glutaraldehyde fixative of high osmolarity for use in electron microscopy. *J. Cell Biol.* **27**, 137A.

LUCKE, B. (1940). The living cell as an osmotic system and its permeability to water. In: *Cold Spring Harbor Symp. Quant. Biol.* **8**, 123–32.

MALHOTRA, S. K. (1962). Experiments on fixation for electron microscopy. I. Unbuffered osmium tetroxide. *Quart. J. micr. Sci.* **103**, 5–15.

MAUNSBACH, A. B. (1966). The influence of different fixatives and fixation methods on the ultrastructure of rat kidney proximal tubule cells. II. Effect of varying osmolality, ionic strength, buffer system and fixative concentration of glutaraldehyde solution. *J. Ultrastruct. Res.* **15**, 283–309.

MILLONIG, G. (1968). In Millonig, G. and Marinozzi, V. Fixation and embedding in electron microscopy. In: *Advances in optical and electron microscopy* (eds. V. E. Coslett & R. Barer). Vol. 2, p. 251. Academic Press: New York.

OLSSON, R. (1961). The skin of amphioxus. *Z. Zellforsch.* **64**, 90–104.

OVALLE, W. K. (1970). Fine structure of rat intrafusal muscle fibres. The polar region. *J. Cell Biol.* **51**, 83.

PERRACHIA, C. (1970). A system of parallel septa in crayfish nerve fibers. *J. Cell Biol.* **44**, 125–33.

PONDER, E. (1940). The red cell as an osmometer. In: *Cold Spring Harbor Symp. Quant. Biol.* **8**, 133–43.

REYNOLDS, E. S. (1963). The use of lead citrate at high pH as an electron opaque stain in electron microscopy. *J. Cell Biol.* **17**, 208–12.

ROBERTSON, E. A. & SCHULTZ, R. L. (1970). The impurities in commercial glutaraldehyde and their effect on the fixation of brain. *J. Ultrastruct. Res.* **30**, 275–87.

SJÖSTRAND, F. S. (1956). Electron microscopy of cells and tissues. In: *Physical Techniques in Biological Research* (eds. G. Oster & A. W. Pollister). Vol. 3, pp. 241–97. Academic Press: New York.

TILNEY, L. G. & GODDARD, J. (1970). Nucleating sites for the assembly of cytoplasmic microtubules in the ectodermal cells of blastulae of *Arbacia punctulata*. *J. Cell Biol.* **46**, 564–75.

YOUNG, J. Z. (1935). Osmotic pressure of fixing solutions. *Nature, Lond.* **135**, 823–5.

The effect of fixative tonicity on the myosin filament lattice volume of frog muscle fixed following exposure to normal or hypertonic Ringer

D. F. DAVEY*

Department of Zoology,
University of Bristol,
UK

Synopsis. Frog sartorius muscles have been fixed sequentially with acrolein and osmium tetroxide dissolved in vehicles of various tonicities, and the myosin filament spacings and sarcomere lengths measured with the electron microscope. From these dimensions the myosin unit-cell volume has been calculated and compared with X-ray diffraction data to determine the effect of fixation. In muscles soaked in normal Ringer and afterwards fixed using normal Ringer as a vehicle for the fixation agents, the unit-cell volume undergoes a 10.4% reduction during the preparative procedure. Muscles soaked in hypertonic Ringer undergo a similar reduction in volume during fixation, provided hypertonic Ringer is used as the vehicle; if they are fixed in normal Ringer, the lattice swells during fixation, even if the change to the normal tonicity vehicle occurs after acrolein fixation. If blocks suitable for embedding are cut from the muscles before, rather than after, osmium fixation, more complex changes in intracellular dimensions may occur, including artefactual swelling of the T-system. It is concluded that fixation of tissues exposed to modifications of normal physiological solutions should be performed using the same modified solutions as fixative vehicles.

Introduction

In any electron microscopic study aimed at determining the effect of modifications of extracellular fluid upon intracellular dimensions, the microscopist is faced with serious problems of experimental design and interpretation. Apart from the usual difficulties of relating dimensions in fixed cells to those in living tissue, preparation of experimentally-modified cells for electron microscopy necessarily involves concentration gradients or transients during fixation different from those invoked during the preparation of the control tissue. Depending upon the nature of the test environment, the microscopist may have to choose the point in the fixation process at which the different transients occur,

* *Present address:* Department of Physiology, Monash University, Clayton, Victoria 3168, Australia.

hopefully in a way to minimize their influence. Then, if examination of the fixed control and experimental tissue reveals an effect, he must decide whether this was brought about by the test modification of the extracellular fluid, or the resultant differences in the fixation transients. In separating the facts from the artefacts, the microscopist all too often has no independent means of making the same measurement.

The experiments reported in this paper were designed to evaluate different approaches to the general problem of fixation of tissue exposed to modified environments through investigation of a simple problem of this type: the effect upon ultrastructure of an increase in the osmotic pressure of the extracellular fluid. The frog sartorius muscle was studied, since the effect of alterations in the osmotic pressure upon the myosin filament array, the major ultrastructural feature of skeletal muscle, has been determined reliably in living fibres by Rome (1968) using X-ray diffraction. Dimensions of the filament lattice in fixed fibres can, therefore, be compared to the dimensions which were found in living fibres subjected to the same experimental conditions, thereby providing an indication of the effectiveness of different fixation methods in preserving the dimensions of the living cells and in preserving the changes produced by the experimental conditions.

The number of combinations of fixatives, concentrations and vehicles that could be considered in a study of this type is virtually unbounded. This investigation has been limited in scope by considering only the effect of the fixative vehicle tonicity on the results obtained from muscles exposed to hypertonic solutions, with a particular view to determining the stability of the fibres following aldehyde fixation. One fixation procedure has been employed throughout the study, and the effect upon the myosin lattice dimensions of this whole fixation embedding and sectioning process has been determined in control tissue. Primary fixation has been carried out with purified acrolein because, first, interesting results have been obtained with this fixative before (Birks & Davey, 1969, 1972); second, its effect on membrane permeability is smaller than that of other fixatives (Carstensen *et al.*, 1971); and last, it penetrates rapidly and produces only slight shrinkage (Luft, 1959).

This study examines three essentially different approaches to the question of whether or not to adjust the fixative vehicle tonicity used in the fixation of the fibres exposed to the hypertonic test solutions. In the first, the vehicles for both the aldehyde and the osmium tetroxide were made hypertonic; in the second, neither was made hypertonic; and in the third only the acrolein vehicle had an elevated tonicity. In addition, because some rather subtle differences in results had been noticed in preliminary experiments, one other small variation was examined: the effect of whether small blocks suitable for embedding were cut from the muscles before or after osmication. The results that follow show that only if the tonicity of both the aldehyde and osmium vehicle is adapted to the test solution is the myosin filament lattice volume appropriately reduced in fixed hypertonic fibres. Better preservation of the myosin lattice volume results if blocks for embedding are cut after osmication, and certain artefacts may be introduced if they are cut out before.

Methods

Solutions

The normal Ringer solution (denoted as *NR*) contained: 115 mM NaCl, 2.5 mM KCl and

1.8 mM $CaCl_2$, as used in previous studies, but the usual phosphate buffer was replaced by 5 mM Hepes (*N*-2-hydroxyethyl-piperazine-*N'*-2-ethanesulphonic acid; as previously employed by Birks, 1971, and Birks & Davey, 1972) and the pH adjusted to 7.0 with NaOH (Good *et al.*, 1966). A Ringer solution of approximately twice normal osmotic pressure (denoted as 2*R*) was prepared by adding 115 mM NaCl to *NR*. No particular significance should be ascribed to the use of Hepes buffer, especially when the Ringer solution is used as a fixative vehicle. Hepes is simply an easy to use, impermeant compound which, unlike tris, buffers well on both sides of neutrality ($pK_{a2} = 7.55$ at 20°C).

Experimental procedure

Sartorius muscles from *Rana pipiens* were excised and mounted spirally and approximately at rest length on 5 mm diameter perspex rods. They were then placed in the test solutions for 10 min at room temperature and fixed on the rods.

Fixation

Acrolein (BDH Ltd) was distilled to eliminate the hydroquinone stabilizer (Carstensen *et al.*, 1971), and was dissolved (2% v/v) in either *NR* or 2*R* as indicated in the text. The use of Ringer solution as a fixative vehicle has several advantages over the use of an isotonic phosphate or cacodylate buffer solution (cf. Brandt *et al.*, 1968; Huxley, 1968). It minimizes transients upon introduction of the fixative; it does not lead to chloride withdrawal contractures (Foulks *et al.*, 1965) at the onset of fixation; its use simplifies the problem of what vehicle to use in preparing fixatives for muscles exposed to modified Ringer solutions, and it is easily adapted to the use of high concentrations of Mg^{++} ions (see Birks, 1971).

Fixation was carried out for 30 min at 4°C, after which the muscles were given three rinses of at least 5 min each in the same Ringer solution. In most experiments small blocks suitable for embedding were cut from one edge of each muscle during the rinse period. Tissue soaked and fixed in 2*R* and which was to be osmicated in *NR* was transferred to *NR* shortly before the onset of osmium fixation. The tissue (including blocks where relevant) was then post-fixed for 1 hr at 4°C in osmium tetroxide (1% w/v) dissolved in buffer-free *NR* or 2*R* as indicated in the text, and then rinsed in several changes of veronal-acetate buffer (Palade, 1952) or maleate buffer (Karnovsky, 1967) at room temperature. If some blocks had been cut before osmication, blocks were now cut from the other edge of the muscle and treated separately; otherwise blocks were cut from one edge only. All blocks were then stained for 45 min at room temperature in uranyl acetate (0.5% w/v) in either veronal-acetate (Kellenberger *et al.*, 1958), or maleate buffer (Karnovsky, 1967) corresponding to the rinse used, then dehydrated in graded ethanol and embedded in Spurr (1969) resin.

Microtomy

The accurate determination of longitudinal and transverse dimensions of fixed muscles using thin sections of embedded material requires that the sections are truly parallel or perpendicular to the fibre longitudinal axis, and that the dimensions of interest are not altered by the sectioning process. The first requirement can be met by careful block orientation, but the second necessitates taking account of the changes in dimensions the sectioned material may undergo during sectioning, usually through the effect known as compression. Compression is the reduction in length of the sectioned material (i.e., parallel to the direction of cutting) in the absence of a change in width (i.e., the dimension parallel to the knife edge), accompanied by a resultant increase in the section thickness. Sections cut from the ERL-embedded blocks used in this study showed compression effects, and comparison of section and block-face dimensions confirmed that the width of the section did not change during cutting (cf. Page & Huxley, 1963). The measurements detailed below were, therefore, always made across the width of the section to avoid the distortion along the length.

Longitudinal sections exhibiting silver-grey interference colours (approx. 550 Å thick; Peachey, 1958) were cut using a Cambridge University Engineering Laboratories—A. F. Huxley pattern microtome equipped with a diamond knife with distilled water in the trough. Blocks were oriented so that the fibres were parallel to the knife edge (cf. Page & Huxley, 1963), and cutting speeds were kept below 0.5 mm/sec. The sections were flattened with chloroform vapour, picked up on collodion-carbon coated grids, stained with lead (Reynolds, 1963), and checked to ensure that they were parallel to the fibre axes. Acceptable grids of a number of specimens were accumulated so that they could be photographed in one session.

Most cross sections were cut, mounted and stained in a manner similar to that described above for longitudinal sections, although some sections were cut on an LKB Ultratome I with glass knives at a cutting speed of 1 mm/sec. Each block was trimmed to include 15–20 fibres; the direction of sectioning was chosen only on the basis of the shape of the resulting block. The sections were checked under the electron microscope to ensure that they were reasonably perpendicular to the fibres, and acceptable grids were accumulated for photography.

Microscope calibration

Microscopes were calibrated using a diffraction grating replica which was itself calibrated using a light microscope. The effect upon magnification of changes in focal length due to differing specimen heights was determined by photographing the replica at a number of specimen heights and recording the objective lens settings. Each microscope was used at only one magnification, and this setting was not altered during calibration or photographic sessions.

Microscopy

Photography of longitudinal sections was carried out with an A.E.I. EM 6B electron microscope set at × 4000. Sections were rotated until the fibres were parallel to one edge of the photographic plate. Three or four micrographs were obtained for each block. The objective lens settings were recorded for each plate so that its true magnification could be determined from the calibration data.

Cross sections were photographed using an A.E.I. EM 6B electron microscope. One micrograph of a region of section passing through the H-zone was obtained from each of 6–10 fibres in each block. Regions with knife marks were occasionally selected to aid orientation of the prints for measurement. The objective lens settings were recorded for each micrograph.

Measurements

Photographic enlargements (approx. $\times$ 2.5) of micrographs of longitudinal sections were measured; the exact magnification of each group of prints in the direction of the fibre axes was determined by direct comparison of the photographic plates and glazed prints. Fibrils which were clearly within the plane of section for the width of the print were selected and the total length of 7–10 sarcomeres was measured. The mean sarcomere length for each measured fibril was treated as one observation for further calculations. The results are presented as the mean $\pm$ S.D.

Approximately 3.5$\times$ photographic enlargements of micrographs of cross sections were used for measurements. Myosin filament separation distances in the $\bar{1},\bar{0}$ lattice plane (see Huxley, 1957) were determined in fibrils in which this plane was accurately parallel to the knife edge, by measuring the total distance along a row of about ten filaments. (The requirement for precise orientation severely restricted the number of observations.) The mean separation between filaments, $d_{\text{m-m}}$, in each measured row was used to calculate the separation distance of the lattice planes, d (1,0) (see April *et al.*, 1971 for diagram), using the relation $d\ (1,0) = d_{\text{m-m}} \sin 60°$. Each result from one row was treated as one observation in further calculations. The results are presented as the mean $\pm$ S.D.

Results

MUSCLES SOAKED IN NORMAL RINGER

Observations on fibres intact and restrained during osmication

Six muscles soaked in *NR* were fixed in acrolein dissolved in *NR* followed by osmium tetroxide in *NR*. The observations on blocks cut from the intact fibres after osmication are given in the upper half of Table 1. As no special care was taken to ensure that the muscles were mounted at a particular length, the data encompass a range of sarcomere lengths, s. Correspondingly the lattice separations, d (1,0), encompass a range such that the volume occupied by one filament, V, referred to as the unit-cell volume (April *et al.*, 1971) or lattice volume (Rome, 1968) and given by $V = [d(1,0)]^2 s \,/\, \sin 60°$, is approximately the same for each muscle (see Table 1). This result demonstrates that the isovolumic behaviour of the frog myofibril reported by others (e.g., Brandt *et al.*, 1967; Rome, 1968) is preserved by acrolein fixation, as can be seen in Fig. 1. The axes of this and subsequent graphs of myosin lattice dimensions are arranged so that isovolumic lines are straight. Line B of Fig. 1 is the isovolumic line for $V = 3.17 \times 10^9$ Å^3, the mean of the observed values (see Table 1), and the points clearly fit this line well.

Despite the fact that the lattice volumes of the fixed fibres confirm the isovolumic behaviour, the mean volume is only 89.6% of that observed in living fibres by Rome ($V_x = 3.54 \times 10^9$ Å^3, line A in Fig. 1). Presumably this discrepancy results from shrinkage during the histological procedures. In confirmation of Luft's (1959) early report that acrolein produces less shrinkage than many fixatives, this shrinkage is less than that observed by Brandt *et al.* (1967) using glutaraldehyde fixation, but agrees with the report of April *et al.* (1971) of 11% shrinkage of the lattice volume during fixation with osmium tetroxide. The latter result is surprising in that most authors have reported problems of swelling during fixation with osmium tetroxide (see Hayat, 1970, Chapter 1 for review).

In view of this apparent agreement between the results obtained with acrolein and osmium tetroxide, it is important to note that the method used to measure the lattice spacings in this report differs significantly from that employed by April *et al.* (1971) as a consequence of different approaches to dealing with the problem of section compression during microtomy. True compression (see *Methods*) should not affect the width of a section, but will reduce its length and area. Consequently, in this study all measurements of filament spacing were made parallel to the section width to avoid any underestimate of myosin spacing due to compression (cf. Carslen *et al.*, 1961). However, April *et al.* choose to average spacings in all three lattice planes, arguing that this would minimize the influence of the compression evident in their micrographs. This argument only holds if the area of the fibre sectioned is conserved during sectioning, otherwise it can lead to a substantial underestimation of lattice spacing, quantitatively dependent upon the extent of the compression. Assuming that the sections of April *et al.* were subject to the typical 20–30% reduction in length, it may well be that their fibres underwent slight swelling during osmium fixation.

The results obtained with *NR*-treated muscles fixed in *NR* appear consistent, and the 10.4% shrinkage will be used as a baseline for comparison in subsequent sections. In

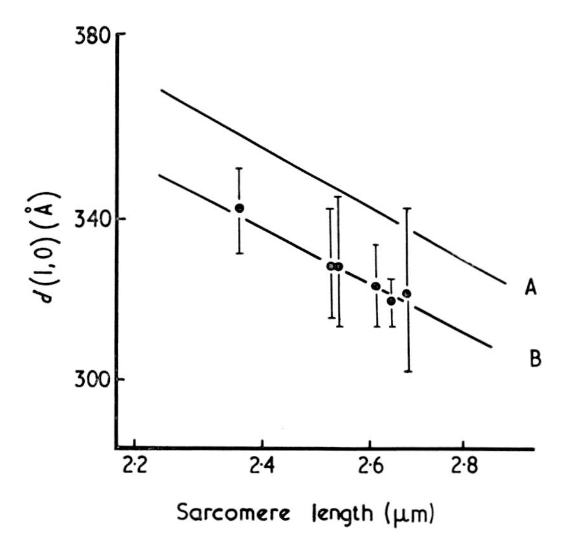

Figure 1. The relationship between the myosin lattice plane separation and sarcomere length in *NR*-treated muscles, observed with the electron microscope after fixation sequentially with 2% acrolein in *NR* and 1% osmium tetroxide in *NR*. The axes were chosen to make isovolumic behaviour of the unit-cell, $[d(1,0)]^2.s$ / sin 60°, appear as straight lines: ordinate, myosin lattice separation, $d(1,0)$, linear with $[d(1,0)]^2$; abscissa, sarcomere length, s, linear with $1/s$. Each point (●) represents the observations on one frog sartorius muscle (see Table 1) and the vertical bars indicate S.D. of $d(1,0)$. The S.D. of the sarcomere lengths averaged about 0.4 μm, but these have not been indicated to retain clarity. Line B is the isovolumic line for unit-cell volume $= 3.17 \times 10^9$ Å³, the mean value observed in the experiments reported in this study. Since the points fit this line, fixation must have preserved the isovolumic behaviour of the myofibrils. However, line A represents the isovolumic behaviour observed in living fibres by Rome (1968) using X-ray diffraction (unit-cell volume $= 3.54 \times 10^9$ Å³), and the points clearly lie below it, presumably because the lattice shrinks during the histological procedures required to examine it in the electron microscope.

Table 1. Lattice dimensions of normal Ringer-treated muscles fixed in normal Ringer.

Muscle	*Blocks cut**	*Sarcomere length,* s† (μm)	*Lattice plane separation* $d(1,0)$† (Å)	*Unit cell volume,* V_{EM}‡ ($\times 10^9$ Å^3)	V_{EM}/V_X§
1	After osmication	2.36 ± 0.01 (22)	342 ± 10 (22)	3.19	0.90
2	After osmication	2.52 ± 0.03 (11)	329 ± 13 (28)	3.15	0.89
3	After osmication	2.53 ± 0.07 (24)	329 ± 16 (22)	3.20	0.84
4	After osmication	2.61 ± 0.04 (21)	324 ± 10 (20)	3.17	0.90
5	After osmication	2.64 ± 0.08 (22)	320 ± 6 (20)	3.12	0.88
6	After osmication	2.67 ± 0.03 (19)	322 ± 20 (31)	3.20	0.90
			Average	3.17 ± 0.03	0.896
1	Before osmication	2.31 ± 0.05 (24)	334 ± 10 (32)	2.98	0.84
2	Before osmication	2.37 ± 0.02 (19)	319 ± 13 (22)	2.78	0.79
3	Before osmication	2.40 ± 0.04 (27)	316 ± 9 (25)	2.76	0.78
6	Before osmication	2.62 ± 0.09 (14)	315 ± 17 (38)	3.00	0.85
			Average	2.88 ± 0.13	0.814

* Indicates whether blocks for embedding were cut before or after osmication.
† Expressed as the mean ± S.D. Bracketed figure is number of measurements. Each measurement is itself an average (see *Methods*).
‡ Unit cell volume, V_{EM}, is given by $[d(1,0)]^2 s/\sin 60°$.
§ V_X is the unit cell volume determined by X-ray diffraction by Rome (1968).
$V_X = 3.54 \times 10^9$ Å^3 for *NR* and 2.44×10^9 Å^3 for 2*R*.

view of the consistency, it is attractive to suggest that the shrinkage may be brought about by a transient osmotic effect of the acrolein in the fixative solution, even though acrolein appears to permeate biological membranes rapidly (Luft, 1959; Griffin, 1963). Unfortunately, the measurements performed in this study do not provide any information concerning when during the fixation procedure the shrinkage occurs. Indeed, compensatory shrinkage and swelling could have occurred at different points during the procedure and not have been detected. Consequently, no attempt has been made to compensate for the shrinkage by reducing the osmotic pressure of the fixative vehicle as proposed by Bone & Denton (1971), although it should be possible to do so by making use of the data presented here together with the X-ray data of Rome (1968). In the absence of other data, there is no assurance that adjustment of the fixative vehicle so as to improve the correspondence between the fixed and living lattice dimensions might not introduce artefacts in the dimensions of other organelles. The question of interest in the remainder of this report is how to fix muscles exposed to solutions other than *NR* and obtain a similar relationship between the observed and expected lattice dimensions.

Observations on fibres cut into short lengths before osmication

In some experiments, blocks were cut from the muscles between the acrolein and osmium fixation stages. The fibres in these blocks differed from those in the part of the muscle that remained intact, in that during osmication they were incomplete, open at each end, and unrestrained. The results obtained from these blocks are given in the second part of Table 1 and the unit-cell volumes are shown in Fig. 2 as open circles. It can be seen from Fig. 2 that the sarcomere lengths and lattice spacings in the blocks are consistently smaller than in the fibres from the same muscles that had been osmicated intact. Consequently, the shrinkage of the myosin lattice is substantially greater in these fibres (averaging 18.6%) and it is also notable that the observed unit-cell volumes

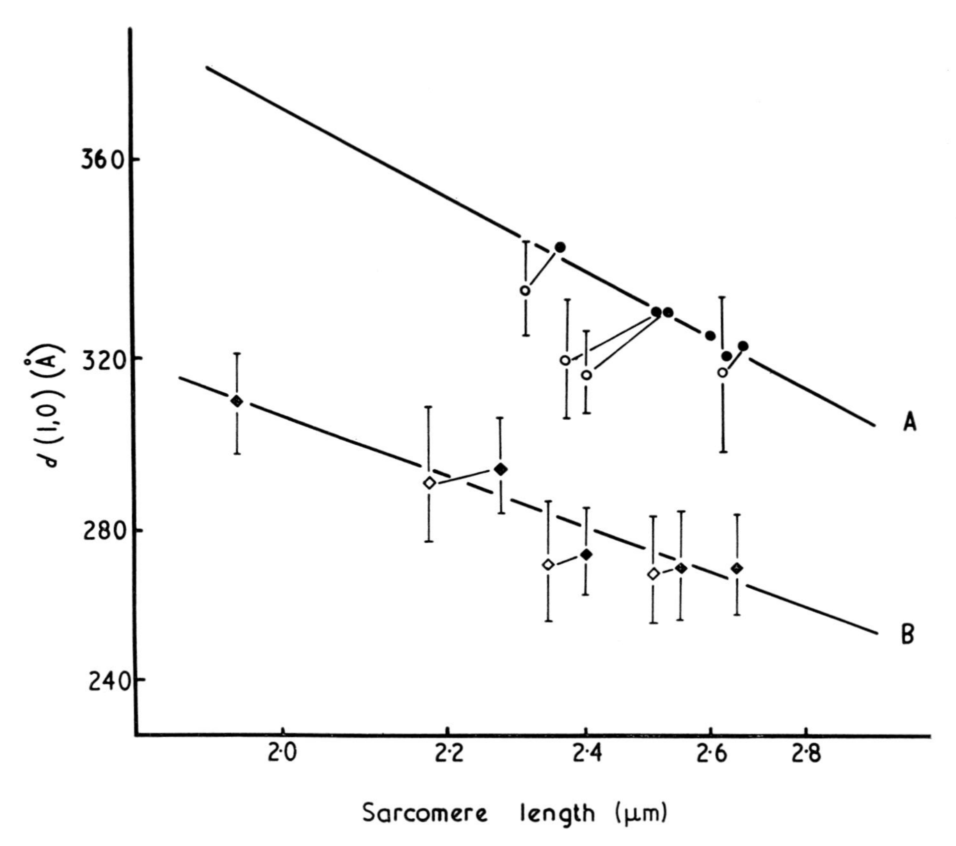

Figure 2. Unit-cell volume in *NR*- and 2*R*-treated muscles fixed in the same Ringer solutions. Line A corresponds to line B in Fig. 1 and the closed circles (●) are the same as those in Fig. 1 with the S.D. bars omitted. The open circles (○) represent the unit-cell volumes from segments of fibres cut from some of the same muscles after acrolein fixation, but before osmication. These cut fibres have undergone more shrinkage than fibres from the same muscle which were intact during osmication. Lines connect open and closed points obtained from the same muscles. Closed and open diamonds represent in the same way, data from 2*R*-treated muscles fixed with acrolein and osmicated as either intact fibres (◆) or cut segments (◇). Line B is the isovolumic line for a unit-cell volume of 2.18×10^9 Å^3, the volume expected if the lattice in living 2*R*-treated muscles undergoes the same 10.4% shrinkage observed in *NR*-treated muscles. Since the points fit this line, it is evident that similar shrinkage does occur and that the isovolumic behaviour of the hypertonic muscles is also preserved by fixation. Axes as in Fig. 1.

encompass a greater range. Both these factors make this variation of the fixation procedure less suitable for obtaining dimensional data.

It seems reasonable to presume that the reduction in sarcomere length resulted from the removal of the longitudinal restraint, but whether the shortening resulted from relaxation of elastic forces present in the acrolein-fixed fibres, or was due to the influence of the osmium tetroxide fixative in the absence of restraint, cannot be determined from the results. Page & Huxley (1963) observed a reduction in filament length in unrestrained fibres fixed with osmium tetroxide alone, and it may be that pre-fixation with acrolein does not prevent this effect. It seems unlikely, however, that the absence of restraint during osmication would result in a decrease in the lattice spacing as well. This effect of cutting blocks before osmium tetroxide fixation is more probably brought about through the associated absence of an intact sarcolemma during osmication, which could produce conditions in the intracellular space of the cut blocks during fixation quite different from those in the intact fibres. Some of these differences (the concentration of fixative, various ions and solutes, and the ionic strength) could influence the lattice dimensions during osmium tetroxide fixation. Regardless of the origin of the reduction in the lattice dimensions that result from this small change in the post-acrolein procedure, its occurrence makes it clear that it is unsafe to assume that all the shrinkage observed when intact fibres are osmicated necessarily occurs during the acrolein fixation stage.

MUSCLES SOAKED IN 2R AND FIXED IN 2R

Table 2 gives the results obtained from 2R-treated muscles fixed sequentially with acrolein and osmium tetroxide dissolved in 2R. The mean unit-cell volume of fibres in blocks cut from the muscles after osmication was 2.18 × 10⁹ Å³, a 31% reduction from

Table 2. Lattice dimensions of 2R-treated muscles fixed in 2R.

Muscle	*Blocks cut**	*Sarcomere length, s*† (μm)	*Lattice plane separation, d*(1,0)† (Å)	*Unit cell volume, V_{EM}*‡ (× 10⁹ Å³)	V_{EM}/V_X§
1	After osmication	1.95 ± 0.03 (15)	309 ± 11 (25)	2.15	0.88
2	After osmication	2.27 ± 0.05 (19)	295 ± 11 (20)	2.29	0.94
3	After osmication	2.40 ± 0.03 (28)	274 ± 11 (24)	2.08	0.85
4	After osmication	2.56 ± 0.15 (15)	270 ± 14 (30)	2.16	0.88
5	After osmication	2.66 ± 0.07 (22)	270 ± 13 (28)	2.24	0.92
			Average	2.18 ± 0.08	0.895
2	Before osmication	2.18 ± 0.05 (17)	293 ± 16 (15)	2.16	0.86
3	Before osmication	2.34 ± 0.02 (14)	272 ± 16 (17)	2.00	0.82
4	Before osmication	2.51 ± 0.02 (16)	269 ± 14 (31)	2.10	0.86
			Average	2.08 ± 0.09	0.856

Footnotes: see Table 1.

the volume determined in *NR*-treated muscles, which is in close agreement with the reduction observed by Rome (1968) with the X-ray diffractometer. However, as was the case in the *NR*-treated muscles, the lattice volume of the fixed fibres is 10.5% less than the volume observed by Rome.

Fig. 2 shows that the unit-cell volume of the 2*R*-treated muscles displays the same isovolumic behaviour seen in the *NR*-soaked fibres, and that the blocks cut from the fibres before osmication (open diamonds) underwent more shrinkage than the corresponding fibres which were intact during osmication (closed diamonds), again analogous to the results obtained with *NR*-treated muscles. Thus it appears that fixation of hypertonic muscles in hypertonic Ringer is a successful approach to the problem of preserving the reduction of the unit-cell volume during the fixation process. Furthermore, the following sections show that it is the only method of those investigated in this study which does preserve this reduction.

MUSCLES SOAKED IN 2*R* AND FIXED IN *NR*

An alternative approach to the preservation of hypertonic muscles is to fix them in the same fixative used for normal tissue, and it can be argued that control and experimental tissue can only be accurately compared if the same fixation procedure has been used for both. Indeed, Huxley *et al.* (1963), in a study of the effects of hypertonic solutions on frog muscle using osmium tetroxide fixation, found that the tonicity of the vehicle did not affect their observations of swellings in the triad system. It is important to note, however, that their measurements indicated a 15% decrease of the spacing between the myosin filaments of 3.5 *R*-treated muscles fixed in normal tonicity osmium tetroxide, whereas extrapolation of the X-ray data of Rome (1968) suggests that this reduction should be about 25%. Similarly, Birks & Davey (1969) found a 4–5% decrease in the myosin filament spacing in 1.5 *R*-treated muscles fixed with acrolein in a normal tonicity vehicle, whereas an 11% decrease is indicated by the X-ray data. This discrepancy is confirmed by the data in Table 3 (plotted in Fig. 3 as upright triangles) which shows that

Table 3. Lattice dimensions of 2*R*-soaked muscles fixed in *NR*.

Muscle	*Blocks cut**	*Sarcomere length, s*† (μm)	*Lattice plane separation, d*(1,0)† (Å)	*Unit cell volume,* V_{EM}‡ (10^9 Å^3)	V_{EM}/V_X§
1	After osmication	2.32 ± 0.06 (15)	314 ± 15 (42)	2.64	1.08
2	After osmication	2.42 ± 0.04 (36)	317 ± 17 (25)	2.81	1.15
3	After osmication	2.47 ± 0.01 (32)	318 ± 10 (34)	2.89	1.18
			Average	2.78 ± 0.13	1.139
1	Before osmication	2.24 ± 0.06 (15)	308 ± 9 (25)	2.45	1.01
3	Before osmication	2.29 ± 0.02 (22)	317 ± 13 (25)	2.65	1.09
			Average	2.55	1.05

Footnotes: see Table 1.

the myosin unit-cell volume of $2R$-soaked muscles is only 12% less than that observed in NR-treated muscles, whereas the X-ray data show that the decrease is 31%.

It is, therefore, clear that to preserve the effects of the soak in hypertonic solutions it is necessary to maintain the elevation in osmotic pressure when fixation is begun. If the osmotic pressure is lowered simultaneously with the introduction of the fixative, the net result will depend upon the relative rates of fixation and re-establishment of osmotic balance. Evidently the myosin lattice is 'fixed' before its volume can completely return to normal in such circumstances, and some reduction in lattice volume is observed. It is not surprising, however, that the range of values obtained is high, and that the points do not show the isovolumic behaviour clearly (see Fig. 3).

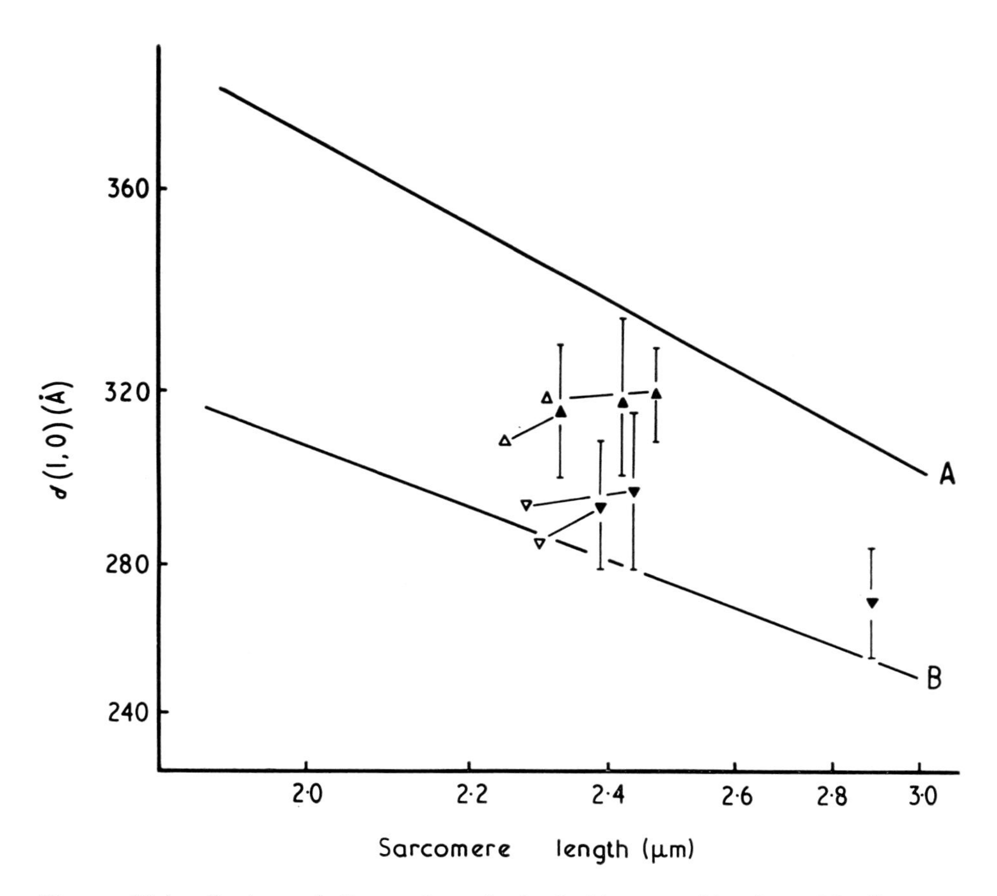

Figure 3. Unit-cell volume of $2R$-treated muscles fixed without, or with only partial, adjustment of the tonicity of the fixative vehicles. Upright triangles: muscles fixed sequentially in acrolein and osmium tetroxide in NR. Inverted triangles: muscles fixed with acrolein in $2R$, but then with osmium tetroxide in NR. Lines joining points relate data obtained from blocks cut before (△,▽) and after (▲,▼) osmication from different fibres in the same muscle. Vertical bars represent S.D., which for clarity have been omitted from the open points. The axes and lines are as in Fig. 2. Note that the data does not fit line B, which is the case when hypertonic muscles are fixed in hypertonic fixatives (see Fig. 2).

MUSCLES SOAKED IN $2R$ AND FIXED WITH ACROLEIN IN $2R$ FOLLOWED BY OSMIUM TETROXIDE IN NR

The results outlined so far have demonstrated the need to match the osmotic pressure of the fixative vehicle to that of the test solution in order to preserve the unit-cell volume, but they do not indicate whether it is essential to adjust both the aldehyde and the osmium tetroxide vehicle, or whether adjustment of the aldehyde vehicle is adequate. It is frequently assumed that this latter alternative is the case, and some experimenters have used the appropriate Ringer solution as an aldehyde vehicle, followed by a standard formulation such as Palade's (1952) osmium tetroxide for post-osmication (see e.g., Brandt *et al.*, 1968). Whether or not such a procedure is valid depends essentially upon whether or not the aldehyde-fixed cells retain osmotic sensitivity; for if aldehyde fixation does not eliminate membrane semipermeability, and Bone & Ryan (1972) have shown that axons are still sensitive to osmotic stress after glutaraldehyde fixation, then a change of osmotic pressure following aldehyde fixation would be expected to alter the volume of partially fixed cells.

This expectation is realized in the case of muscle fibres test-soaked and acrolein-fixed in $2R$, then transferred to NR just before osmication in NR (see Table 4). The mean unit-cell volume is less than that observed if fixation is in NR throughout, but greater than that obtained with fixation in $2R$ throughout. Hence the fibres have undergone an osmotic response after acrolein fixation, but neither to the degree nor with the scatter observed when the osmotic step occurs at the onset of acrolein fixation.

Although the behaviour of the unit-cell volume seen with this fixation procedure does not coincide with the behaviour observed when the osmium vehicle is also hypertonic (Fig. 3), the isovolumic behaviour is still apparent. Evidently the acrolein fixed fibres are still capable of undergoing an osmotic response (as would be expected in view of the observations of Bone & Ryan, 1972), but the acrolein fixation has simply reduced the magnitude of this response, at least in so far as the myosin lattice is concerned. Clearly

Table 4. Lattice dimensions of $2R$-treated muscles fixed in acrolein in $2R$ followed by osmium tetroxide in NR.

Muscle	*Blocks cut**	*Sarcomere length, s*† (μm)	*Lattice plane separation* $d(1,0)$† (Å)	*Unit cell volume* V_{EM}‡ (10^9 Å^3)	V_{EM}/V_X§
1	After osmication	2.38 ± 0.06 (15)	293 ± 15 (27)	2.36	0.97
2	After osmication	2.43 ± 0.06 (22)	296 ± 18 (24)	2.46	1.01
3	After osmication	2.89 ± 0.02 (33)	270 ± 13 (19)	2.43	1.00
			Average	2.41 ± 0.05	0.991
1	Before osmication	2.29 ± 0.10 (9)	285 ± 12 (29)	2.15	0.88
2	Before osmication	2.27 ± 0.03 (28)	294 ± 13 (28)	2.27	0.93
			Average	2.21	0.91

Footnotes: see Table 1.

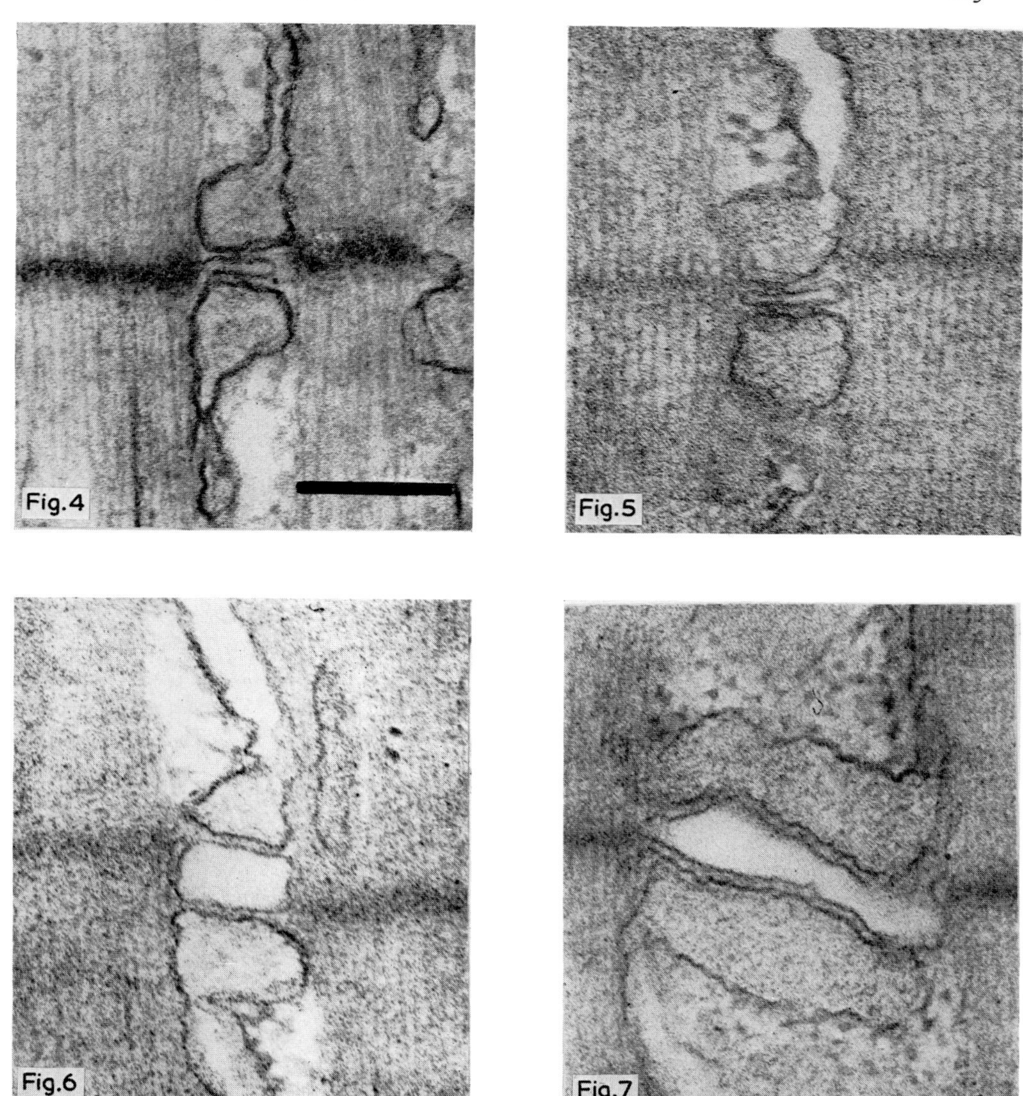

Figure 4. Longitudinal section of a normal Ringer-soaked frog sartorius fibre fixed sequentially with acrolein and osmium tetroxide dissolved in *NR* showing a transverse tubule in cross section (centre). Scale bar = 0.2 μm.

Figure 5. Similar view to Fig. 4 but from a hypertonic Ringer-soaked muscle fixed with acrolein in hypertonic Ringer but then osmicated using *NR* as a vehicle. Note that the transverse tubule appears normal despite the reduction in osmotic pressure associated with the change in vehicle.

Figures 6 & 7. Longitudinal sections from fibres treated in the same way as the one shown in Fig. 5 except that blocks suitable for embedding were cut from the acrolein-fixed muscles before osmium tetroxide fixation. The fact that the fibres were not intact during osmication resulted in swelling of the transverse tubules during fixation, presumably because of osmotic gradients set up across the tubule membranes by the reduction in vehicle tonicity.

the volume increase of the fibres as a whole may be greater than the lattice volume increase suggests, for the fixed lattice probably resists osmotic forces upon it owing to cross-linking of the filaments.

It is interesting that in experiments in which blocks were cut from the muscles before the fixative vehicle was changed from 2*R* to *NR* (Table 4, second group) the resulting lattice volumes are very close to those observed where no such osmotic step occurred, for this suggests that the osmotic adjustments of the fixed lattice are partly dependent upon the integrity of the sarcolemma. In these blocks an observation was made which underlines the importance of the osmotic balance of the fixative vehicle, and the failure of the unit-cell volume alone to provide an index of preservation quality. Parts of the transverse tubular system (T-system) of fibres cut open before transfer from 2*R* to *NR* underwent substantial swelling (Figs. 4–7), even though, as stated above, the myosin lattice did not swell significantly. Presumably, following the change of vehicle, the intracellular osmotic pressure falls rapidly in the fibre segments because the ends are not membrane bounded, whereas the T-system lumen might still contain hypertonic Ringer, especially if acrolein fixation has restricted the patency of these tubes or if they have been disrupted by fixation into isolated segments, thus eliminating the *in vivo* continuity with the extracellular compartment. If membrane semipermeability is still present, as the results of Bone & Ryan (1972) suggest to be the case, osmotic swelling of these T-segments would follow. This possibility may be of significance in studies of T-system volume, and its relevance to certain work in the literature will be considered in the discussion.

Discussion

The results obtained in this study show that after sequential fixation with acrolein and osmium tetroxide in normal Ringer and dehydration and embedding in plastic, the unit-cell volume of normal Ringer-soaked frog sartorius muscles is reduced by this preparative procedure to about 90% of the volume observed in living fibres with X-ray diffraction. This shrinkage appears consistent, regardless of the sarcomere length, and the results confirm the isovolumic behaviour of the lattice volume. The unit-cell volume of muscles soaked in 2*R* is observed to undergo a similar reduction to 90% of the volume in living fibres, provided the fixation procedure is carried out in 2*R*. If *NR* is used as the vehicle in fixative solutions applied to 2*R*-soaked muscles, the myosin lattice swells during fixation. Furthermore, the results show that the acrolein-fixed fibres can undergo an osmotic response. Consequently, the reduction in lattice volume produced by hypertonic Ringer solution is preserved only if hypertonic fixative vehicle is used right through to the osmium tetroxide fixation stage. It seems, therefore, that if cellular dimensions of fibres exposed to anisotonic solutions are to be preserved, fixative solutions prepared with the same anisotonic solutions should be used.

This conclusion reflects the fact that fixation does not occur instantaneously. If the introduction of the fixative solution involves a change in the composition of the extracellular fluid that before fixation would lead to a redistribution of solutes or water, there is a strong possibility that these gradients may effect the same changes to the ultrastructure before the fixation process has stabilized it. The major factor determining whether such gradients will affect ultrastructure is the speed with which the fixation agent eliminates semi-permeability; for once this occurs, osmotic gradients should no longer affect the volume of the membrane bound compartments, whether they be cells or organelles.

There is considerable evidence indicating that when osmium tetroxide is applied to living cells, membrane permeability increases rapidly (Elbers, 1966; Millonig, 1968), and consequently it has been found that the tonicity of the vehicle used for osmium tetroxide is of little importance (Huxley *et al.*, 1963; Wood & Luft, 1965; Elbers, 1966; Rapoport *et al.*, 1969). It has been tempting to suppose that aldehyde fixatives might produce similar permeability changes, either upon their introduction, or with a longer time-course as fixation ensued, so that the osmolarity or at least the composition of the aldehyde fixative vehicles would also not be exceptionally important. Investigations into the validity of this supposition have not entirely clarified the issue; in the erythrocyte, aldehyde fixation appears to eliminate osmotic responsiveness (Carstensen *et al.*, 1971), whereas it does not in the sea urchin egg (Millonig, 1968), in the fish scale (Bone & Denton, 1971) or in the crab axon (Bone & Ryan, 1972).

This apparent contradiction raises the question of whether or not the osmotic behaviour of fixed cells accurately reflects the permeability properties of the fixed membrane. If it does, one could conclude that fixation of the erythrocyte membrane eliminated its semipermeability, whereas in other tissues it was maintained. However, it is difficult to separate the properties of the fixed membrane from the properties of the fixed cell as a whole; the cellular contents may become rigidly precipitated during fixation, and the membrane may be bound to these fixed cellular contents.

The experiments reported here show that 2*R*-soaked muscles fixed in acrolein and finally osmicated in *NR* display different myosin lattice volumes dependent upon whether the acrolein fixation was carried out in *NR* or 2*R*. Since the results obtained at the end of these two alternative osmotic pathways are not only dependent on the initial and final osmolarity, acrolein must reduce the responsiveness of the myosin lattice by making it more rigid, and the final volume of the unit cell depends upon whether the change from the hypertonic vehicle occurs at the onset, or following, acrolein fixation.

If this is true for cells generally, then the osmotic responsiveness of a fixed cell should depend not only upon the properties of the membranes, but on the properties of the fixed cellular constituents. If a living cell contains a sufficient density of material, the fixed cytoplasmic matrix may be sufficiently rigid to enable the cell to resist osmotic shock, which may well explain why Carstensen *et al.* observed that the red blood cell, with its dense cytoplasm, is not sensitive to osmotic shock after fixation. On the other hand, the responsiveness of material with a relatively low concentration of material in the cytoplasm may be relatively unaffected by fixation. Clearly, the observed responsiveness of the fixed myosin lattice lies somewhere between these two extremes, and does not necessarily indicate that fibre semipermeability is altered by acrolein fixation.

In fact, it is evident that even osmium tetroxide also does not immediately eliminate the osmotic responsiveness of the acrolein-fixed fibres. Otherwise the swelling of fibres soaked and acrolein-fixed in 2*R* that resulted when the osmolarity of the bathing medium was decreased virtually at the onset of osmium tetroxide fixation should not have been observed. This result is in accord with the observation of Elbers (1966) who noted that a dramatic increase in permeability of the *Limnaea* egg occurs with osmium tetroxide fixation, but that after fixation with glutaraldehyde, osmium tetroxide causes only a much slower increase in permeability.

There is a slight difficulty of interpretation associated with this point in the present study, for the change from $2R$ to NR occurred just before osmication, not simultaneously with the introduction of the osmium fixative. Consequently, it could be argued that the osmotic response was complete before the osmium fixative was applied. However, in experiments conducted with Professor R. I. Birks, in which the change in tonicity occurred precisely with the onset of osmium fixation, the same results were obtained (Birks & Davey, 1969, and unpublished observations).

It seems reasonable to conclude from these considerations that membrane semipermeability is not significantly altered by aldehyde fixation. Consequently, if one is interested in the effects of a modified physiological solution on the ultrastructure of a cell, the solution should be present *during* the aldehyde fixation (and probably during the early stages of osmium fixation as well) in order to maintain the gradients until the ultrastructure has been completely fixed. If this is not done, the structure of the fixed tissue may well represent the structure brought about by the *effux* of the test solution.

That problems can arise if osmotic gradients are established during fixation is particularly evident in the experiments in which small fibre segments of hypertonic Ringer-soaked muscles were subjected to a rise in osmotic pressure after acrolein fixation. The T-system of these blocks underwent substantial swelling, presumably because the membranes were still semipermeable at the time of the osmotic pressure change; the only pathway for the solutes in the T-system to leave was through the openings to extracellular space. On the other hand, the osmolarity change would be rapidly transmitted to the intra-fibre space because the ends of the fibres were cut open and an osmotic gradient could be established resulting in water movement into the T-tube lumen before osmium tetroxide fixation was completed. The degree of this artefactual swelling could be sensitive to differences in the composition of the fluid in the T-tubes at the time of the osmolarity increase. The less permeant the solutes in the T-system lumen are, the more effective in producing artefactual swelling they should be. Consequently, T-system swelling in tissue handled in this way (see e.g., Brandt *et al.*, 1968) could be an assay for changes in composition of the T-tubular fluid, not an indication of tubular changes occurring in the living fibres.

A similar mechanism could be responsible for the T-system swelling seen in muscles exposed to solutions containing sucrose before fixation with osmium tetroxide alone (Freygang *et al.*, 1964, 1967; Rapoport *et al.*, 1969) and interpreted as an artefact by Birks & Davey (1972). It may be that as the T-system membranes become more permeable during osmication, there is a period during which the membranes become more permeable to small ions, but have not yet become permeable to sucrose; during such a period, the T-system would undergo osmotic swelling, perhaps just at a time when final stabilization of the membranes is taking place.

Thus it appears that the safest way of preserving tissue exposed to modified physiological solutions is to fix it sequentially with aldehyde and osmium tetroxide, and the best vehicle for the fixing agents appears to be the same physiological solution bathing the tissue before fixation. Although this process may be longer, and subject to the need for greater attention to the nature of the vehicles than is the use of osmium tetroxide alone, artefacts are less likely to result (cf. Birks & Davey, 1972). Where the removal of small tissue blocks may cut open cells of interest, this cutting should be left until after osmication, otherwise the overall shrinkage may be increased, the dimensions may be less consistent, and strong gradients not present in life may be established directly across the membranes of incompletely fixed organelles.

Acknowledgements

This study was supported by the Muscular Dystrophy Association of Canada. The author is indebted to Dr P. C. Caldwell and Dr A. E. Dorey for the provision of facilities and for helpful suggestions during the course of this study and the preparation of the manuscript. Assistance of Mr J. Clement (E.M. maintenance), Mr D. King (photography), and Miss G. McCrindle (manuscript preparation) is gratefully acknowledged.

References

APRIL, E. W., BRANDT, P. W. & ELLIOT, G. F. (1971). The myofilament lattice: studies on isolated fibers. I. The constancy of the unit-cell volume with variation in sarcomere length in a lattice in which the thick-to-thin myofilament ratio is 6:1. *J. Cell. Biol.* **51**, 72–82.

BIRKS. R. I. (1971). Effects of stimulation on synaptic vesicles in sympathetic ganglia, as shown by fixation in the presence of Mg^{2+}. *J. Physiol., Lond.* **216**, 26–28P.

BIRKS, R. I. & DAVEY, D. F. (1969). Osmotic responses demonstrating the extracellular character of the sarcoplasmic reticulum. *J. Physiol., Lond.* **202**, 171–88.

BIRKS, R. I. & DAVEY, D. F. (1972). An analysis of volume changes in the T-tubes of frog skeletal muscle exposed to sucrose. *J. Physiol., Lond.* **222**, 95–111.

BONE, Q. & DENTON, E. J. (1971). The osmotic effects of electron microscope fixatives. *J. Cell Biol.* **49**, 571–81.

BONE, Q. & RYAN, K. P. (1972). Osmolarity of osmium tetroxide and glutaraldehyde fixatives. *Histochem. J.* **4**, 331–47.

BRANDT, P. W., LOPEZ, E., RUEBEN, J. P. & GRUNDFEST, H. (1967). The relationship between myofilament packing density and sarcomere length in frog striated muscle. *J. Cell Biol.* **33**, 255–63.

BRANDT, P. W., REUBEN, J. P. & GRUNDFEST, H. (1968). Correlated morphological and physiological studies on isolated single muscle fibers. II. The properties of the crayfish transverse tubular system: localization of the sites of reversible swelling. *J. Cell Biol.* **38**, 115–29.

CARSLEN, F., KNAPPEIS, G. G. & BUCHTAL, F. (1961). Ultrastructure of the resting and contracted striated muscle fibers at different degrees of stretch. *J. biophys. Cytol.*, **11**, 95–117.

CARSTENSEN, E. L., ALDRIDGE, W. G., CHILD, S. Z., SULLIVAN, P. & BROWN, H. H. (1971). Stability of cells fixed with glutaraldehyde and acrolein. *J. Cell Biol.* **50**, 529–33.

ELBERS, P. F. (1966). Ion permeability of the egg of *Limnaea stagnalis* L., on fixation for electron microscopy. *Biochim. biophys. Acta* **112**, 318–29.

FOULKS, J. G., PACEY, J. A. & PERRY, FLORENCE A. (1965). Contractures and swelling of the transverse tubules during chloride withdrawal in frog skeletal muscle. *J. Physiol., Lond.* **180** 96–115.

FREYGANG, W. H., GOLDSTEIN, D. A., HELLAM, D. C. & PEACHEY, L. D. (1964). The relation between the late after-potential and the size of the transverse tubular system of frog muscle. *J. gen. Physiol.* **48**, 235–63.

FREYGANG, W. H., RAPOPORT, S. I. & PEACHEY, L. D. (1967). Some relations between changes in the linear electrical properties of striated muscle fibers and changes in ultrastructure. *J. gen. Physiol.* **50**, 2437–58.

GOOD, N. E., WINGET, G. D., WINTER, WILHELMINA, CONNOLLY, T. N., IZAWA, S. & SINGH, R. M. M. (1966). Hydrogen ion buffers for biological reseaıch. *Biochemistry, N.Y.* **5**, 467–77.

GRIFFIN, J. L. (1963). Motion picture analysis of fixation for electron microscopy: *Amoeba proteus*. *J. Cell Biol.* **19**, 77A.

HAYAT, M. A. (1970). *Principles and Techniques of Electron Microscopy*. Volume I, Biological applications. New York: Van Nostrand Reinhold.

HUXLEY, H. E. (1957). The double array of filaments in cross-striated muscle. *J. biophys. biochem. Cytol.* **3**, 631–48.

HUXLEY, H. E. (1968). Structural difference between resting and rigor muscle: evidence from intensity changes in the low-angle equatorial X-ray diagram. *J. molec. Biol.* **37**, 507–20.

HUXLEY, H. E., PAGE, SALLY & WILKIE, D. R. (1963). An electron microscopic study of muscle in hypertonic solutions. Appendix to Dydyńska, Maria & Wilkie, D. R. *J. Physiol., Lond.* **169**, 312–29.

KARNOVSKY, M. J. (1967). The ultrastructural basis of capillary permeability studied with peroxidase as a tracer. *J. Cell Biol.* **35**, 213–36.

KELLENBERGER, E., RYTER, ANTOINETTE & SÉCHAUD, JANINE (1958). Electron microscope study of DNA-containing plasms. II. Vegetative and mature phage DNA as compared with normal bacterial nucleoids in different physiological states. *J. biophys. biochem. Cytol.* **4**, 671–8.

LUFT, J. H. (1959). The use of acrolein as a fixative for light and electron microscopy. *Anat. Rec.* **133**, 305.

MILLONIG, G. (1968). Experimental approach to problems of fixation and dehydration. *Advances in Optical and Electron Microscopy* **2**, 287–317.

PAGE, SALLY G. & HUXLEY, H. E. (1963). Filament lengths in striated muscle. *J. Cell Biol.* **19**, 369–90.

PALADE, G. E. (1952). A study of fixation for electron microscopy. *J. exp. Med.* **95**, 285–98.

PEACHEY, L. D. (1958). Thin sections. I. A study of section thickness and physical distortion produced during microtomy. *J. biophys. biochem. Cytol.* **4**, 233–42.

RAPOPORT, S. I., PEACHEY, L. D. & GOLDSTEIN, D. A. (1969). Swelling of the transverse tubular system in frog sartorius. *J. gen. Physiol.* **54**, 166–77.

REYNOLDS, E. S. (1963). The use of lead citrate at high pH as an electron-opaque stain in electron microscopy. *J. Cell Biol.* **17**, 208–12.

ROME, ELIZABETH (1968). X-ray diffraction studies of the filament lattice of striated muscle in various bathing media. *J. molec. Biol.* **37**, 331–44.

SPURR, A. R. (1969). A low-viscosity epoxy resin embedding medium for electron microscopy. *J. Ultrastruct. Res.* **26**, 31–43.

WOOD, R. L. & LUFT, J. H. (1965). The influence of buffer systems on fixation with osmium tetroxide. *J. Ulstrastruct. Res.* **12**, 22–45.

The demonstration of acid phosphatase in in vitro *cultured tissue cells. Studies on the significance of fixation, tonicity and permeability.*

ULF T. BRUNK
and JAN L. E. ERICSSON

Department of Pathology, University of Uppsala, Uppsala; and Department of Pathology at Sabbatsberg's Hospital, Karolinska Institutet Medical School, Stockholm, Sweden

Synopsis. 1. Methods for the fine structural demonstration of acid phosphatase were studied in monolayers of *in vitro* cultured cells after fixation with glutaraldehyde.

2. Inactivation of enzyme activity occurred rapidly during the initial phase of glutaraldehyde fixation.

3. Fixation for more than 5 min did not cause further marked inactivation of enzyme activity.

4. Stabilization of the cells for cytochemical incubations required a fixation for at least 30 min in glutaraldehyde.

5. The total osmolality of the fixative was of minor importance, in contrast to the major importance of effective osmolality, for obtaining optimum cytochemical and ultrastructural results.

6. Following proper fixation, the osmotic strength of the washing and incubation solutions was not critical.

7. With short fixation times, the composition of the washing and incubation solutions was of major importance.

8. Dimethyl sulphoxide in washing and incubation media was effective in shortening incubation times (thereby preventing the occurrence of unspecific precipitates and derangement of fine structure).

Introduction

The bulk of previous histochemical studies on the localization of acid phosphatase in different cell types have been performed on cells *in situ*. All the different preparatory procedures utilized for such studies have been analysed with regard to the effects of the composition of the fixative and washing solutions, as well as the incubation media, on the fine structure of the cells and the preciseness and reproducibility of the deposition of the final reaction product. These investigations have led to improvements in the histochemical techniques, thus permitting the precise localization of acid phosphatase in structurally well-preserved cells (Beck & Lloyd, 1969; Holt & Hicks, 1962; Reale & Luciano, 1970; Sabatini *et al.*, 1963; Trump & Ericsson, 1965).

In the case of cells in tissue culture, detailed analyses of the effects on the fine structure and enzyme localization of the different preparatory procedures used for the visualization of acid phosphatase are, however, largely lacking.

Preliminary experiments revealed that experience from studies on whole tissues are not in all facets applicable to single cells in tissue culture when attempting to demonstrate acid phosphatase in such cells whilst preserving their proper morphology. The aim of the present paper is to analyse some of the factors that influence the results of demonstrating acid phosphatase cytochemically in *in vitro* cultured cells at the light and electron microscope levels.

Materials and methods

Culture conditions

Normal human glia cells and embryonic rat fibroblasts were utilized for the investigation. Human diploid glia cell cultures were maintained and continuously propagated with the incubation and subculture techniques described previously (Pontén *et al.*, 1969). The embryonic rat fibroblasts were obtained from minced Sprague-Dawley rat embryos and cultivated in the same manner as the human cells. All lines were propagated in plastic Falcon or Nunc dishes and only early passages were used.

Fixation and washing

When the cells were ready for fixation the tissue culture dishes were removed from the incubation chambers, the Eagle's medium removed and substituted with fixative at 0°C, and the fixation performed with the dishes kept on melting ice. The fixation times were varied within broad limits, ranging from 15 min to 24 hr.

Glutaraldehyde, purified by distillation and charcoal washing (Anderson, 1967), was used in concentrations ranging between 2 and 4.5%. Apart from the glutaraldehyde, the composition of the fixative solution was varied with respect to the content of sodium cacodylate-HCl buffer (pH 7.2) and sucrose: the concentration of buffer was 0, 0.1 or 0.15 M and sucrose 0, 0.1, 0.2 or 0.3 M. After fixation had been completed, the fixative solution was substituted with a washing solution (at 0°C) that was changed three times during the first ten min; the wash was continued at 0°C for periods ranging between 1 and 24 hr. The washing solutions were either 0.1 or 0.15 M sodium cacodylate-HCl buffer (pH 7.2) with or without sucrose added to a concentration of 0.1 or 0.2 M, or 0.25 M sucrose containing 0.01 M Tris-HCl buffer (pH 7.2) with or without 10 or 20% v/v dimethyl sulphoxide (DMSO) added.

The osmotic pressures of the different solutions were measured with a Fiske model G osmometer. Selected examples of the osmolalities of the different solutions used for fixation, washing and incubation are given in Table 4.

Biochemical assays

For the assessment of the inactivation of acid phosphatase during fixation and washing, a pool of trypsinized human glia cells was prepared in Eagle's minimal essential medium. This suspension of cells was then equally divided between the plastic Falcon dishes (diameter 100 mm) used for the experiment, thus making sure that each dish contained about the same number of cells. The cells were then cultivated until a confluent monolayer showing contact inhibition of movement was formed, *i.e.* for 5 days. The number of cells in the different dishes—after establishment of a confluent monolayer—varies with less than 5% when this method for subcultivation is used (Westermark, personal communication, 1972). The biochemical assays were performed on cells fixed for 5 min to 24 hr in 2 or 3% glutaraldehyde in 0.1 M sodium cacodylate-HCl buffer (pH 7.2) containing 0.1 M sucrose. They were washed in 0.25 M sucrose in 0.01 M Tris-HCl buffer (pH 7.2) or in the same washing solution containing 10% DMSO. The washing time was varied between 1 and 24 hr. Unfixed cells rapidly rinsed in saline served as controls.

After fixation and washing, the cells were rapidly rinsed in glass-distilled water, scraped off in 2.00 ml glass-distilled water, homogenized in a Potter-Elvehjem homogenizer equipped with a glass pestle, and stored in a deep freeze (at −70°C) until the assays were performed.

Incubations for measurements of total activity of acid phosphatase were performed as described by Bowers *et al.* (1967). To ensure complete rupture of the lysosomal membrane, Triton X-100 was added to the incubation mixture in a final concentration of 0.1% (Bowers *et al.*, 1967). The incubation was performed at 37°C for 60 min at pH 5.0. (In this diluted system the rate of production of hydrolysis product was found to be linear with time for more than 120 min.) Inorganic phosphate was determined as described by Itaya & Ui (1966). The enzyme activity was expressed as the relative ratio of the extinctions of the inorganic phosphate assay solutions to that obtained for control cultures (submitted neither to fixation nor washing).

Cytochemistry

Following the various modes of fixation and washing, the cultures were incubated in a modified Gomori-type medium (Barka & Anderson, 1962) with or without 0.22 M sucrose added or 10% v/v DMSO added or with the addition of both sucrose and DMSO. The incubations were performed with constant agitation at 37°C for times ranging between 15 and 120 min. Before incubation, the culture dishes were rapidly warmed to 37°C by two short rinses with the prewarmed incubation solution. Controls were incubated in either a medium lacking β-glycerophosphate or a complete medium containing 0.01 M NaF. After the incubation, the cultures were rinsed in three changes of ice-cold 0.9% saline.

For light microscopy the reaction product was visualized by immersing the cultures briefly in 1% ammonium sulphide. The cultures were then post-fixed for 60 min in 3% glutaraldehyde in 0.15 M sodium cacodylate-HCl buffer (pH 7.2), rinsed in glass-distilled water, and air dried. They were then mounted under cover glasses with a drop of immersion oil.

For electron microscopy the cells were post-fixed for 90 min at room temperature in 2% osmium tetroxide in *s*-collidine buffer after rinsing in 0.9% saline.

Electron microscopy

A preparatory technique that allowed the analysis of large numbers of cells with preserved shape and relationships to each other and their supporting structure was used. The method, slightly modified from Biberfeld (1968), is described in detail elsewhere (Brunk *et al.*, 1971).

Cells were studied, after the different types of fixation and washing described, either directly or following incubation for the demonstration of acid phosphatase.

Results

Quantitation of enzyme inactivation

A rapid decline in the activity of acid phosphatase was noted following the application of fixative (Fig. 1). Thus, 5 min after exposure of the cells to 2% glutaraldehyde, only about 20% of the initial activity was retained. At later intervals (up to 60 min), a very slight further decrease was recorded. After 24 hr fixation, approximately 10% of the activity

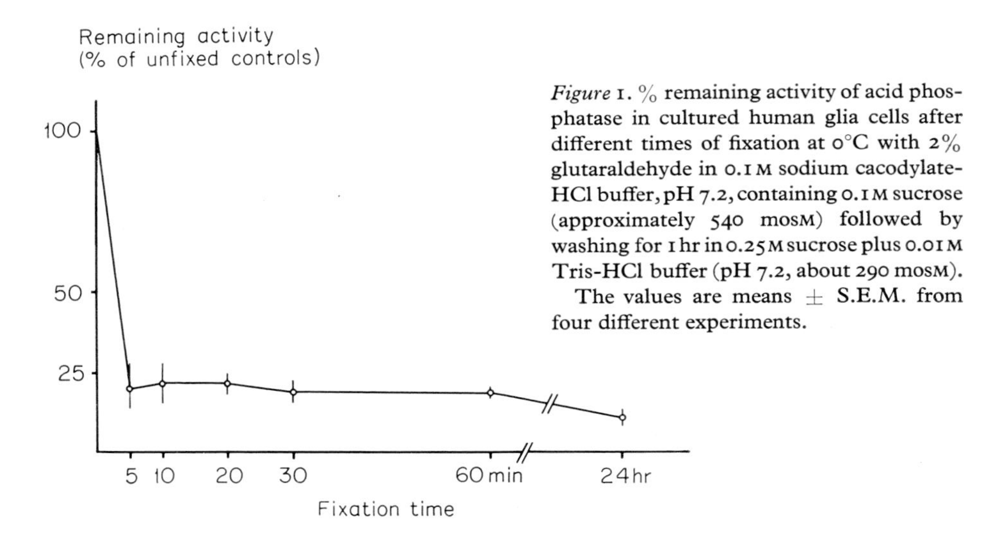

Figure 1. % remaining activity of acid phosphatase in cultured human glia cells after different times of fixation at 0°C with 2% glutaraldehyde in 0.1 M sodium cacodylate-HCl buffer, pH 7.2, containing 0.1 M sucrose (approximately 540 mosM) followed by washing for 1 hr in 0.25 M sucrose plus 0.01 M Tris-HCl buffer (pH 7.2, about 290 mosM).

The values are means ± S.E.M. from four different experiments.

remained. As illustrated in Fig. 2, washing periods varying between 1 and 24 hr did not significantly alter the remaining activity after 30 min fixation of the cells, and neither did the presence of 10% DMSO in the washing solution. No significant difference in the inactivation of enzyme activity was observed between 2% and 3% glutaraldehyde.

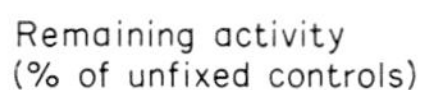

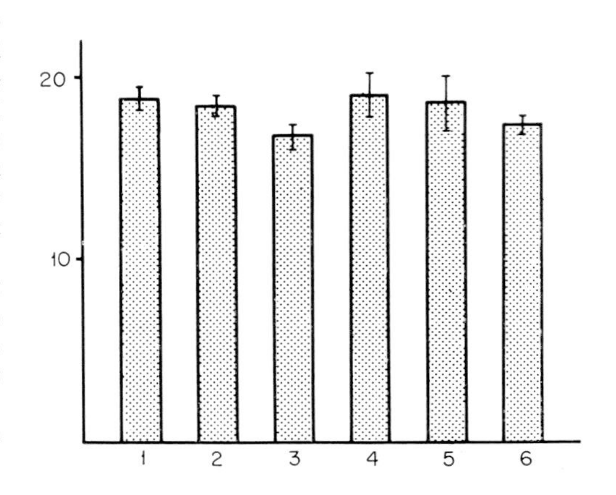

Figure 2. % remaining activity of acid phosphatase in cultured human glia cells fixed for 30 min at 0°C with 2% glutaraldehyde in 0.1 M sodium cacodylate-HCl buffer, pH 7.2, plus 0.1 M sucrose (about 540 mosM) after different times of washing at 0°C with 0.25 M sucrose in 0.01 M Tris-HCl buffer (pH 7.2, about 290 mosM) with or without 10% v/v DMSO added.

Bar 1: No DMSO, 1 hr wash. Bar 2: DMSO added, 1 hr wash. Bar 3: No DMSO, 2 hr wash. Bar 4: No DMSO, 4 hr wash. Bar 5: DMSO added, 4 hr wash. Bar 6: No DMSO, 24 hr wash.

The values are means ± S.E.M. from four different experiments.

Effects of variations in the composition of fixative, washing solution and incubation medium on cytochemically demonstrable acid phosphatase

(*a*) After *brief fixation* (15 min), pronounced variations in the pattern of deposition of the final reaction product was noted, the pattern depending on the composition of the fixative as well as on the washing and incubation solutions. As illustrated in Fig. 3, a distinct granular type of final reaction product was obtained when fixation was performed in 2% glutaraldehyde in 0.1 M sodium cacodylate-HCl buffer plus 0.1 M sucrose (approximately 540 mosM) provided the washing solution and Gomori medium also contained sucrose in amounts giving an osmolality of approximately 300 mosM. Higher concentrations of sucrose in the fixative, washing solution and incubation medium resulted in shrinkage of the cells (see below) and slower appearance of the granular 'staining'. Fig. 4 shows the distribution of final reaction product in cells fixed and further processed in solutions lacking sucrose (and of low osmolality). Enlarged and partly confluent granular sites of final reaction product, slightly reduced in number, and severe diffuse cytoplasmic 'staining' was noted. With appropriate fixation, followed by washing and incubation under low osmolality conditions, some diffuse cytoplasmic deposition of final reaction product occurred while the granular pattern was still apparent (Fig. 5). The findings after different combinations of fixative, washing and incubation procedures are summarized in Table 1.

Table 1. Effects of the composition of the washing solution and the Gomori medium on the 'staining' quality in the cytochemical demonstration of acid phosphatase after a short (15 min) fixation at 0°C with 2% glutaraldehyde (GA) in 0.1 M sodium cacodylate-HCl buffer (Cac) with or without 0.1 M sucrose (Sucr) added (pH 7.2, osmolality about 540 or 430 mosM respectively). Washing was performed at 0°C for 4 hr in 0.1 M sodium cacodylate-HCl buffer, with or without 0.1 M sucrose added (pH 7.2, osmolality about 300 or 200 mosM respectively). The Gomori medium was prepared with and without 0.22 M sucrose (approximately 300 mosM and 50 mosM respectively) and the incubation was run for 90 min at 37°C and pH 5.0. +++ indicates distinct, granular type of 'staining', + indicates indistinct 'staining' with diffuse cytoplasmic precipitation of the final reaction product and/or few granules (cf. Figs. 3 & 4).

Fixative	*Washing Sol.*	*Incubation Sol.*	*'Staining'*
GA + Cac + Sucr	Cac + Sucr	Gomori medium + Sucr	+++
GA + Cac + Sucr	Cac + Sucr	Gomori medium	++(+)
GA + Cac + Sucr	Cac	Gomori medium + Sucr	++
GA + Cac + Sucr	Cac	Gomori medium	++
GA + Cac	Cac + Sucr	Gomori medium + Sucr	+
GA + Cac	Cac + Sucr	Gomori medium	+
GA + Cac	Cac	Gomori medium + Sucr	(+)
GA + Cac	Cac	Gomori medium	(+)

(*b*) After *long fixation* (24 hr) a granular staining pattern was obtained regardless of whether or not sucrose was present in the washing or incubation solutions or both provided the fixative contained both 0.1 M sucrose and 0.1 M cacodylate buffer (Table 2).

Table 2. Effects of the composition of the washing solution and the Gomori medium on the 'staining' quality in the histochemical demonstration of acid phosphatase after a long (24hr) fixation. The fixatives, the washing solutions, the incubation techniques and abbreviation are the same as those given in the legend of Table 1.

Fixative	*Washing Sol.*	*Incubation Sol.*	*'Staining'*
GA + Cac + Sucr	Cac + Sucr	Gomori medium + Sucr	+++
GA + Cac + Sucr	Cac + Sucr	Gomori medium	+++
GA + Cac + Sucr	Cac	Gomori medium + Sucr	+++
GA + Cac + Sucr	Cac	Gomori medium	+++
GA + Cac	Cac + Sucr	Gomori medium + Sucr	+

(*c*) Following fixation in solutions with approximately *equal osmolality* (approximately 500 mosM) the staining pattern differed widely and depended on the composition of the fixative when washing and incubation were performed under ideal conditions (see Table 1). As shown in Fig. 7, fixation in 4.5% glutaraldehyde in 0.01 M 'cacodylate buffer' resulted in diffuse 'staining' of the cytoplasm and the presence of final reaction product in the nuclei. Absence of cacodylate buffer in the fixative (2% glutaraldehyde in 0.3 M sucrose) gave the same type of reaction as when 0.1 M sucrose was used in combination with 0.1 M cacodylate buffer (Fig. 6). Granular 'staining' was also noted following fixation with 2% glutaraldehyde in 0.15 M cacodylate buffer, but the time required to obtain a reaction corresponding to that after fixation in fixative with 0.1 M sucrose and 0.1 M cacodylate buffer was considerably shorter.

Figure 3 (Colour Plate). Pattern of demonstrable acid phosphatase activity in *in vitro* cultured embryonic rat fibroblasts after short fixation (15 min) at 0°C with 2% glutaraldehyde in 0.1 M sodium cacodylate-HCl buffer made 0.1 M with respect to sucrose (pH 7.2, approximately 540 mosM), followed by washing for 4hr at 0°C in 0.1 M sodium cacodylate-HCl buffer with 0.1 M sucrose added (pH 7.2, approximately 320 mosM). The cells were incubated for 90 min at 37°C in modified Gomori medium containing 0.22 M sucrose (pH 5.0, approximately 300 mosM).

Note distinct granularity and absence of diffuse nuclear or cytoplasmic 'staining'. × 600

Figure 4 (Colour Plate). Pattern of demonstrable acid phosphatase activity in *in vitro* cultured embryonic rat fibroblasts after short fixation (15 min) at 0°C in 2% glutaraldehyde in 0.1 M sodium cacodylate-HCl buffer (pH 7.2, approximately 410 mosM, i.e. no sucrose added), followed by washing for 4hr at 0°C in 0.1 M sodium cacodylate-HCl buffer (pH 7.2, about 200 mosM; no sucrose added). The cells are incubated in the modified Gomori medium (pH 5.0, approximately 50 mosM) without sucrose at 37°C for 90 min.

Note prominent diffuse 'staining' of cytoplasm and nucleus, and enlargement and coalescence of many granules. × 600

Figure 5 (Colour Plate). Pattern of demonstrable acid phosphatase activity in *in vitro* cultured embryonic rat fibroblasts after short fixation (15 min) at 0°C in 2% glutaraldehyde in 0.1 M sodium cacodylate-HCl buffer made 0.1 M with respect to sucrose (pH 7.2, approximately 540 mosM), followed by washing for 4hr at 0°C in 0.1 M sodium cacodylate-HCl buffer (pH 7.2, approximately 200 mosM; no sucrose added). The cells are incubated in the modified Gomori medium (pH 5.0, about 50 mosM) without sucrose at 37°C for 90 min. × 600

Note moderate number of distinct granules and rather pronounced diffuse, cytoplasmic 'staining'.

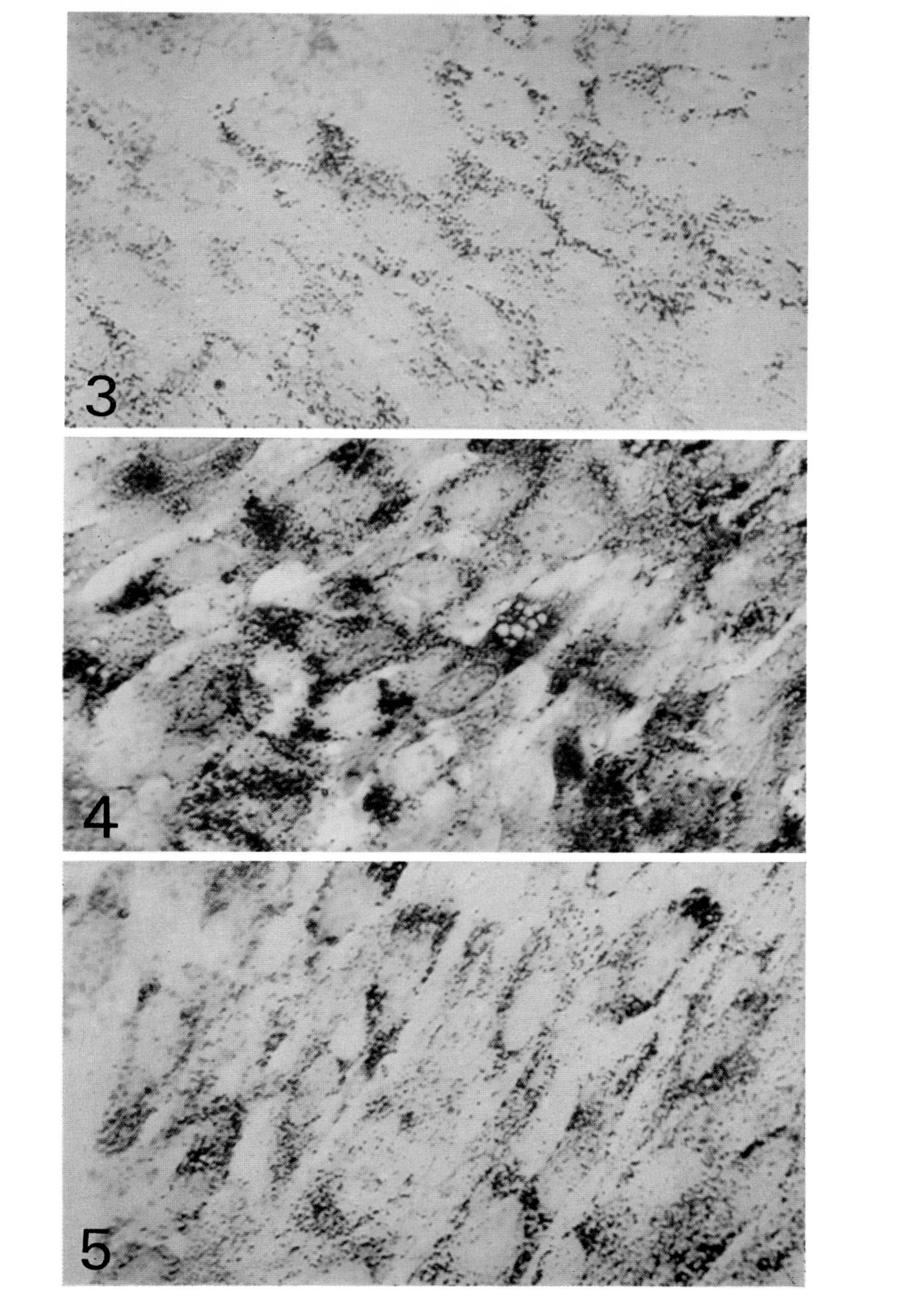
3
4
5

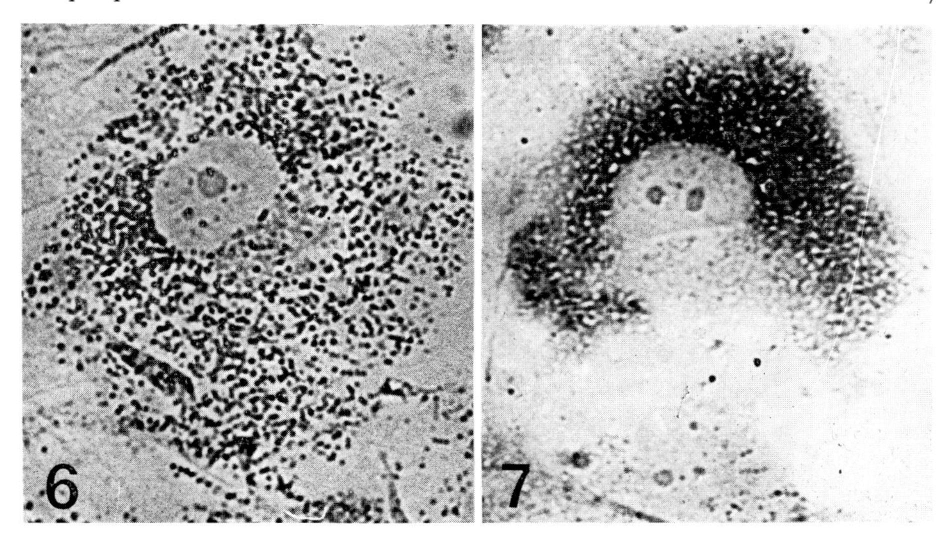

Figure 6. Pattern of demonstrable acid phosphatase activity in *in vitro* cultured human glia cells after fixation for 1 hr at 0°C in 2% glutaraldehyde in 0.1 M sodium cacodylate-HCl buffer made 0.1 M with respect to sucrose (pH 7.2, about 540 mosM), followed by washing for 1 hr at 0°C in 0.25 M sucrose containing 0.01 M Tris-HCl buffer (pH 7.2, about 290 mosM) and incubation for 60 min at 37°C in the modified Gomori medium (pH 5.0, about 1800 mosM) containing 0.22 M sucrose and 10% v/v DMSO.

Note granular type of 'staining'. × 900

Figure 7. Cytochemical procedure as described in legend of Fig. 6, with the modification that fixation was performed in 4.5% glutaraldehyde in 0.01 M sodium cacodylate-HCl buffer (pH 7.2, about 540 mosM).

Note diffuse type of 'staining'. × 900

Effect of DMSO on the histochemical reaction

The time needed to obtain a distinct granular reaction could be considerably shortened by addition of DMSO to the Gomori solution, and even more so by including it both in the washing solution and the Gomori medium (Table 3). The osmolality of the Gomori

Table 3. Effect of DMSO on time required to obtain a distinct, granular 'staining' in glia cells when acid phosphatase was demonstrated histochemically with a modified Gomori medium containing 0.22 M sucrose with or without 10% v/v DMSO added. The cells were fixed for 60 min at 0°C in 2% glutaraldehyde in 0.1 M sodium cacodylate-HCl buffer containing 0.1 M sucrose (pH 7.2), with or without 10% DMSO added.

Washing and incubation	*Incubation time needed to obtain appropriate granular reaction*
No DMSO	60–90 min
DMSO in Gomori medium	30–45 min
DMSO in both washing solution and Gomori medium	20–30 min

solution containing 0.22 M sucrose and 10% DMSO was found to be approximately 1800 mosM, and that of the DMSO containing washing solution about the same.

Remarks on the fine structure obtained after fixation

After brief fixation (15 min) the preservation of the fine structure following incubation in the Gomori medium was poor, even when optimal conditions for fixation, washing and incubation (see Table 1) were employed. When fixation was carried out for 1 hr or longer the preservation was adequate provided the fixative, washing and incubation solutions were those indicated in Table 1 as giving optimum results (Figs. 8 & 9). With a combined

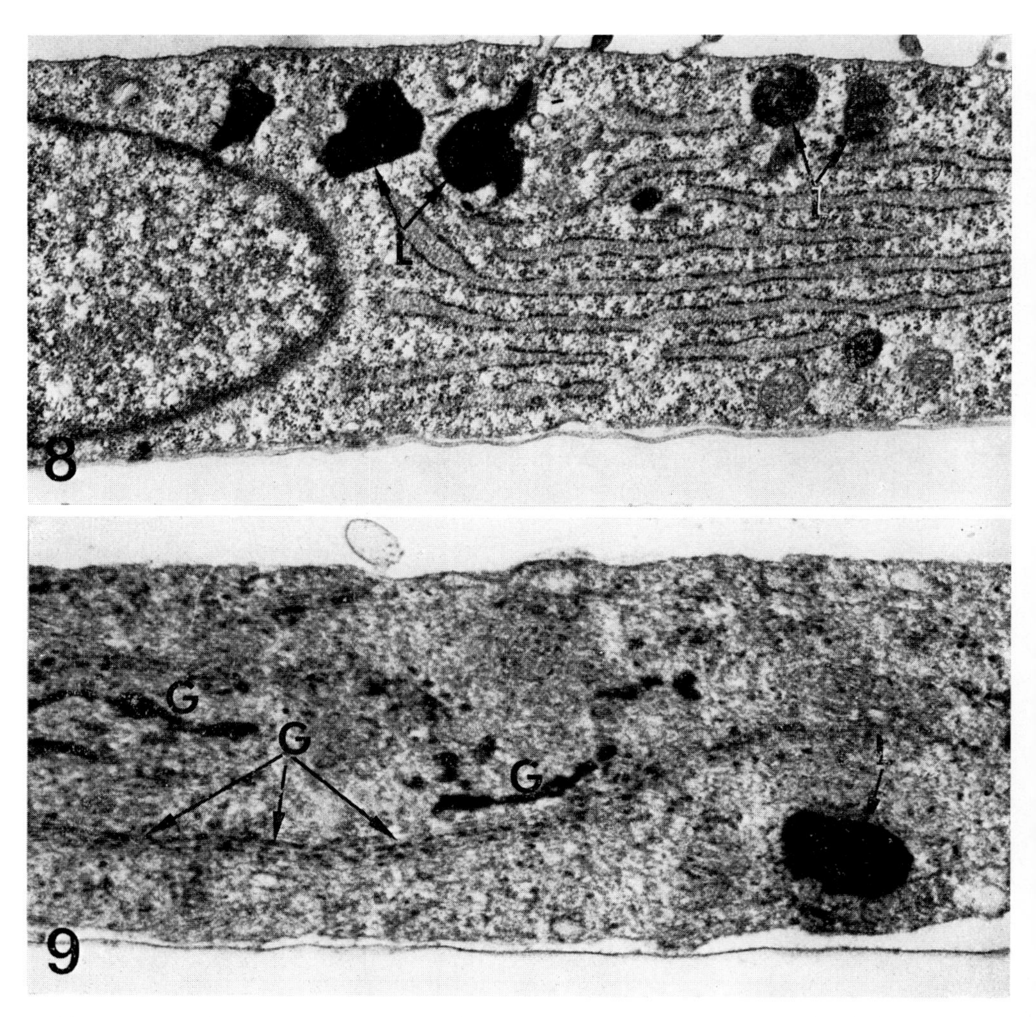

Figures 8 & 9. Electron micrographs of cells incubated for the demonstration of acid phosphatase activity according to the method recommended at the conclusion of the *Discussion*.

Lysosomes (L) as well as Golgi regions (G) containing final reaction product are illustrated. The tissue preservation is only moderately disturbed during the relatively short incubation time (30 min) and the amount of non-specific lead phosphate precipitation is small. Fig. 8, × 20 000; Fig. 9, × 48 000

osmolality of cacodylate and sucrose in the fixative greater than about 300 mosM, shrinkage of the cells was noted as manifested by diffusely increased density of the ground substance or cytoplasm in the cells. The same applies when the combined osmolality of sucrose and cacodylate was increased in the washing solution, or when more than 0.22 M sucrose was added during the incubation in the Gomori medium. However, the addition of 10% DMSO to the washing solution and Gomori medium, which increased the osmolality to about 1800 mosM, had no such dehydrating effect.

Table 4. Osmolality of selected solutions utilized in the different experiments.

Solutions	*Approximate mosM*
Eagle's MEM with 10% calf serum	300
2% GA in 0.1 M cacodylate buffer and 0.1 M sucrose	540
2% GA in 0.15 M cacodylate buffer	550
2% GA in 0.3 M sucrose	530
4.5% GA in 0.01 M cacodylate buffer	540
0.25 M sucrose in 0.01 M Tris-HCl buffer	290
0.1 M sucrose in 0.1 M cacodylate buffer	320
0.1 M cacodylate buffer	200
Gomori medium (Barka & Anderson, 1962)	50
Gomori medium with 0.22 M sucrose	300
Gomori medium with 0.22M sucrose and 10% v/v DMSO	1800

Discussion

The ultimate goal in enzyme cyto- and histochemistry is to demonstrate activity in morphologically well-preserved tissues in sites where the enzyme is normally located and without losing significant amounts of activity during the various preparatory steps which lead to the final assessment of the localization. The usual experience is that one has to compromise between the demands for complete preservation of enzyme activity and structure—at least at the electron microscopic level. Thus, in order to obtain proper preservation of fine structure, fixation is usually needed; this leads to more or less pronounced inactivation of enzyme activity. In the present investigation our main goals were (*a*) to elucidate the degree of inactivation of acid phosphatase during fixation; (*b*) to observe the effects of the composition of the fixative, washing, and incubation solutions on the localization of acid phosphatase and the fine structure of the cells; and (*c*) to study the effects of the permeability-promoting compound DMSO on fine structure, localization of enzyme, and activity of enzyme during the washing and incubation procedures.

It could be expected that the net result of attempts to demonstrate acid phosphatase with a Gomori-type reaction should be more favourable in *in vitro* cultured cells than in sections of solid tissues, the reason being that there are no problems regarding penetration and temperature effects during fixation, at least not when a monolayer culture is used. Further, there is no need for sectioning before incubation is performed. Nevertheless it appeared in preliminary studies of cultured cells (Brunk & Ericsson, 1972) that the results of enzyme cytochemistry, as seen in the light microscope and more especially in the electron microscope, were less successful than was anticipated, and inferior to those obtained with solid material after freezing and sectioning during 'DMSO protection' (Helminen & Ericsson, 1970), or after preparation of non-frozen sections (Frank & Christensen, 1968; Smith & Farquhar, 1966).

The main reason for the inferiority of the results with *in vitro* cultured cells seems to be that rather long incubation times are required to obtain final reaction product in amounts sufficient for its visualization. It should be remembered that an ordinary Gomori-type medium offers a milieu which is by no means suitable for the preservation of fine structure because of its low pH and low osmolality. The long time required for the enzyme cytochemical demonstration seems to be the comparatively high resistance of the membranes to the penetration of the substrate (β-glycerophosphate)—even after aldehyde fixation (Bone & Denton, 1971). When sections of tissues are used, the cell membranes are presumably opened up or made more easily permeable by the cutting or freezing procedures.

Since a prerequisite for further fine structural observations is to achieve a precise localization of the final reaction product, the effects of the various preparatory steps on the outcome of the Gomori reaction were first studied at the light microscope level. The evidence available to date suggests that, under normal conditions in the types of cells studied, acid phosphatase (EC. 3.1.3.2.) is an enzyme exclusively associated with lysosomes, Golgi elements and related structures and does not occur free in the cell sap. Distinct granular and structure-bound localization of the final reaction product in histochemical and cytochemical studies of acid phosphatase can, therefore, be used as an indication of the proper treatment and incubation for the tissues and cells investigated.

In some experiments, spreading of final reaction product to produce a diffuse cytoplasmic and nuclear staining was noticed. It appears likely that this resulted from diffusion of enzyme (and not reaction product) since it was only noted after exposure to solutions with low effective osmotic pressure; the control incubations seem to exclude non-specific binding of lead ions to extralysosomal sites possibly created by the low osmotic treatment. Thus, in the assessment of the results in the present study, diffuse localization of reaction product has been regarded as a sign of inability of the treatment to keep the enzyme immobilized.

As shown in Fig. 1, the inactivation of acid phosphatase is a fairly rapid process which for practical purposes is completed within 5 min when a monolayer of tissue-cultivated cells is utilized. However, a short fixation is not sufficient to immobilize the enzyme, as indicated in Table 1. After a short fixation (15 min) the enzyme is still mobile, and a pronounced diffusion will occur if rinsing and incubation does not take place during carefully controlled conditions (cf. Figs. 3 & 4). On the other hand, a long fixation will result in adequate binding of the enzyme, and the cells will be able to withstand even rather rough rinsing and incubation procedures, as indicated in Table 2. The electron microscopic observations showed that treatments resulting in proper localization of enzyme also yielded adequate preservation of fine structure.

The logical interpretation of these findings is that the fixation time should be long enough to ensure a reliable stabilization of the tissue and that there is no significant gain of enzyme activity by keeping the fixation time short.

It has previously been shown that glutaraldehyde solutions which are distinctly hypertonic to tissue fluid nevertheless can cause tissues to swell (Schultz & Karlsson, 1965; Torack, 1965). Thus, Schultz & Karlsson (1965) found that rat brain tissue fixed by perfusion with glutaraldehyde solution showed hypotonic damage when the total osmolality of the fixative was equal to that of cerebrospinal fluid. However, good results were obtained when the total osmolality was much higher, provided the osmolality of the buffer was about the same as that of the cerebrospinal fluid. Ordinary formaldehyde fixatives used for light microscopy, e.g. 4% formaldehyde in water, have a high osmolality (well above 1000 mosM); nevertheless, after fixation the tissues show hypotonic alterations, and salts have to be added to the fixative to avoid this damage. Baker (1965) found that better fixation for electron microscopy can be obtained if sucrose is added to the formaldehyde fixative to increase the osmotic pressure.

In a recent study on the reflecting cells of fish scales, Bone & Denton (1971) showed that short fixation with aldehyde solutions did not destroy the semipermeability of cellular plasma membranes and that no changes in volume of the cells took place provided the vehicle of the fixative solution had an osmolality of about 60% of that to which the cells were in equilibrium in life.

Our findings indicate that lysosomal acid phosphatase diffuses inside the cells during the process of fixation unless the vehicle for the aldehyde fixative has an osmolality of approximately 300 mosM (or more). This figure is about the same as the osmolality of Eagle's medium used for cultivating the cells. With 2% glutaraldehyde, the total osmolality of the fixative (glutaraldehyde + vehicle) then becomes about 540 mosM. However, when the same osmolality was obtained with higher concentration of glutaraldehyde in a low osmolar vehicle, pronounced enzyme diffusion was noted. This might result from swelling with subsequent rupture of the lysosomes or altered permeability of their membranes.

It seemed to be of some importance whether the osmolality of the vehicle was obtained with sucrose only, sodium cacodylate-HCl buffer only, or combinations of both. If buffer only was used during a short fixation (less than 30 min), the incubation period required to obtain a distinct granular 'staining' with the modified Gomori method was found to be shorter than if either sucrose or sucrose plus buffer was used. It thus seems that the membrane permeability towards β-glycerophosphate was affected by the composition of the fixative vehicle. This is in accordance with the well-known phenomenon that 0.25 M sucrose is a better preservative for isolated lysosomes than salt solutions with the same osmolality. Further, as shown by Berthet *et al.* (1951), different salts have different properties in this respect, acetate ions being more effective than chloride ions.

The total osmotic pressure of the fixative solution seems to be of minor importance as compared to 'the effective osmotic pressure' which might be defined as the pressure of those 'particles' which do not pass the semipermeable membranes under investigation, *e.g.* plasma membranes and lysosomal membranes. Fig. 10 illustrates schematically the differences between total and effective osmotic pressure, and the effects of exposure of membrane-limited structures to solutions of variable osmolalities.

Glutaraldehyde obviously passes through the cellular membranes freely and accordingly does not contribute to the effective osmotic pressure (Bone & Denton, 1971). Sucrose probably does not pass the membranes and thus constitutes one of the compounds in the fixative solution which take part in the building up of the effective osmotic pressure. In living cells also, small molecules like sodium, calcium, and chloride ions contribute to the effective osmotic pressure. In what way they do this during fixation is not known; however, it can be presumed that the functional capacity of the membranes changes when the fixation process has started. For instance, the sodium pump of the membranes will cease to function, and the permeability of the membranes against smaller molecules, like the ones just mentioned, will possibly change during fixation. Therefore, it is difficult to define exactly which particles form the effective osmotic pressure. It is entirely possible that different buffers of the same osmotic pressure provide different effective osmotic pressures. However, as tested in this system, an osmolality of 300 mosM of the fixation vehicle, when made up of 0.1 M sucrose and 0.1 M sodium cacodylate-HCl buffer, gave a satisfactory result.

As shown in Tables 1 & 2, the stability of the cells after different periods of fixation differed considerably. After a short fixation (15 min) the cells are still osmotically active and swelling resulting in diffusion of proteins, including organelle-bound ones like acid phosphatase, readily occurs during rinsing as well as incubation. This is in agreement with results published by Hopwood (1969) and Hopwood *et al.* (1970) that indicate that the process of fixation with aldehydes is a gradual and slow cross-linking of the proteins.

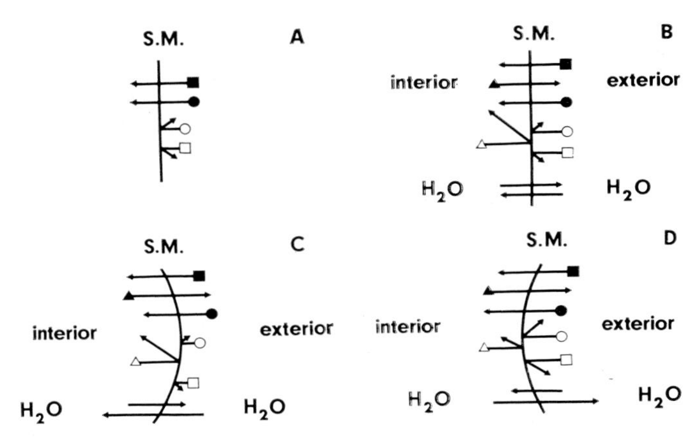

Figure 10. Crude schematic representation of the suggested difference between 'total' and 'effective' osmolality with examples of the results of variability in osmotic effect of molecules and ions which do and do not pass a semipermeable membrane (S.M.). The figures illustrate three tentative examples of net molecular movements when a complex solution surrounds a membrane-limited structure.

A. Illustration of difference between osmotically-active and osmotically-inactive compounds.
Total osmotic pressure: accumulated osmotic pressure of compounds ■, ●, ○, and □ (as measured by freezing-point depression).
Effective osmotic pressure: accumulated osmotic pressure of compounds ○ and □ which do not pass the semipermeable membrane.

B. Effective osmotic pressure or external solution is about the same as that of internal solution.

C. Effective osmotic pressure of external solution is lower than that of internal solution (with resultant net movement of water to the interior = 'swelling').

D. Effective osmotic pressure of external solution is higher than that of internal solution (with resultant net movement of water to the exterior = 'shrinkage' or 'dehydration').
In examples B–D, total osmotic pressure of the external solution is higher than that of the internal solution.

It should be noted that the osmotic pressure of the Gomori-type solution lacking sucrose is only about 50 mosM and, therefore, causes swelling of insufficiently fixed tissue. In contrast, after a long fixation cells are able to withstand even rough treatments as shown by the absence of diffusion of acid phosphatase in sections examined in the light microscope; presumably the cell structure is completely stabilized.

When monolayers of cultured cells are incubated for the demonstration of acid phosphatase, the incubation time needed to obtain a sufficient amount of reaction product is comparatively long. The reason for this appears to be that the membranes still show latency and are not freely permeable to the substrate molecules. A long incubation period evidently has negative effects on the fine structure of the cells, probably because of the low pH of the incubation solution. Furthermore, the long incubation time produces unspecific and unavoidable electron-dense lead precipitates in the tissue that disturb, and sometimes invalidate, the interpretation of possible enzyme reactivity in small intracellular sites, such as vesicles associated with the Golgi apparatus or other organelles.

It is thus of great importance to keep the incubation period as short as possible. As shown in Table 3, DMSO very efficiently diminishes the incubation time needed, especially if it is included both in the washing solution and incubation medium. It is known from studies by others that this effect of DMSO can be taken advantage of in enzyme histochemistry (Gander & Moppert, 1969; Makita & Sandborn, 1971; Misch & Misch, 1969); however, the exact mechanism is not understood. Possibly, water molecules in the membranes are substituted with DMSO which may render the membranes more permeable, perhaps by influencing the size of their pores. It is important in this connection to note that DMSO appears to be freely permeable, and hence does not influence the effective osmolality of the solutions in which it is present.

On the basis of the findings in this study we recommend the following method for the cytochemical demonstration of acid phosphatase in cells cultured *in vitro*: (i) *Fix* at 0°C for 1–24 hr with 2% purified glutaraldehyde in 0.1 M sodium cacodylate-HCl buffer (pH 7.2) containing 0.1 M sucrose (total osmolality approximately 540 mosM); (ii) *wash* for 1–4 hr at 0°C in 0.25 M sucrose containing 0.01 M buffer (sodium cacodylate-HCl or Tris-HCl, pH 7.2) and 10% v/v DMSO (total osmolality about 1800 mosM); and (iii) *incubate* at 37°C in a Gomori-type medium (pH 5.0) made 0.22 M with respect to sucrose and containing 10% v/v DMSO (total osmolality about 1800 mosM).

Acknowledgements

The skilled assistance of Miss Silwa Mengarelli, Mrs Britt-Marie Åkerman, Mr Bengt-Arne Fredriksson, and Mr Magnus Norman is gratefully acknowledged.

This work was supported by grants from the Swedish Medical Research Council (projects B71-12X-1006-06C and B72-12X-1007-07A) and the Swedish Cancer Society (grant number 71:208).

References

ANDERSON, P. J. (1967). Purification and quantitation of glutaraldehyde and its effect on several enzyme activities in skeletal muscle. *J. Histochem. Cytochem.* **15**, 652–61.

BAKER, J. R. (1965). The fine structure produced in cells by fixatives. *J. R. Microsc. Soc.* **84,** 115–31.

BARKA, T. & ANDERSON, P. J. (1962). Histochemical methods for acid phosphatase using hexazonium pararosanilin as a coupler. *J. Histochem. Cytochem.* **10,** 741–53.

BECK, F. & LLOYD, J. B. (1969) Histochemistry and electron microscopy of lysosomes. In *Lysosomes in Biology and Pathology* (eds. J. T. Dingle and H. B. Fell), Vol. 2, pp. 567–99. North Holland: Amsterdam and London.

BERTHET, J., BERTHET, L., APPELMANS, F. & DE DUVE, C. (1951). Tissue fractionation studies. 2. The nature of the linkage between acid phosphatase and mitochondria in rat-liver tissue. *Biochem. J.* **50,** 182–9.

BIBERFELD, P. (1968). A method for the study of monolayer cultures with preserved cell orientation and interrelationship. *J. Ultrastruct. Res.* **25,** 158–9.

BONE, Q. & DENTON, E. J. (1971). The osmotic effects of electron microscope fixatives. *J. Cell Biol.* **49,** 571–81.

BOWERS, W. E., FINKENSTAEDT, J. T. & DE DUVE, C. (1967). Lysosomes in lymphoid tissue. I. The measurement of hydrolytic activities in whole homogenates. *J. Cell Biol.* **32,** 325–37.

BRUNK, U., ERICSSON, J. L. E., PONTÉN, J. & WESTERMARK, B. (1971). Specialization of cell surfaces in contact-inhibited human glia-like cells *in vitro*. *Exptl. Cell Res.* **67,** 407–15.

BRUNK, U. & ERICSSON, J. L. E. (1972). Ultrastructural enzyme histochemistry of cells in tissue culture. Significance of fixation, tonicity and permeability factors, as exemplified in studies of 'acid phosphatase'· *J. Ultrastruct. Res.* **38,** 192.

FRANK, A. L. & CHRISTENSEN, A. K. (1968). Localization of acid phosphatase in lipofuscin granules and possible autophagic vacuoles in interstitial cells of the guinea pig testis. *J. Cell Biol.* **36,** 1–13.

GANDER, E. S. & MOPPERT, J. M. (1969). Der Einfluss von Dimethylsulfoxid auf die Permeabilität der Lysosomenmembran bei quantitativer und qualitativer Darstellung der sauren Phosphatase. *Histochemie* **20,** 211–14.

HELMINEN, H. & ERICSSON, J. L. E. (1970). On the mechanism of lysosomal enzyme secretion. Electron microscopic and histochemical studies on the epithelial cells of the rat's ventral prostate lobe. *J. Ultrastruct. Res.* **33,** 528–49.

HOLT, S. J. & HICKS, M. (1962). Combination of cytochemical staining methods for enzyme localization with electron microscopy. In *The Interpretation of Ultrastructure* (ed. R. J. C. Harris). Symposia of the International Society for Cell Biology, Vol. 1, pp. 193–211. Academic Press: New York and London.

HOPWOOD, D. (1969). Fixatives and fixation: a review. *Histochem. J.* **1,** 323–60.

HOPWOOD, D., ALLEN, C. R. & MCCABE, M. (1970). The reactions between glutaraldehyde and various proteins. An investigation of their kinetics. *Histochem. J.* **2,** 137–50.

ITAYA, K. & UI, M. (1966). A new micromethod for the colorimetric determination of inorganic phosphate. *Clin. Chim. Acta* **14,** 361–6.

MAKITA, T. & SANDBORN, E. B. (1971). The effect of dimethyl sulfoxide (DMSO) in the incubation medium for the cytochemical localization of succinate dehydrogenase. *Histochemie* **26,** 305–10.

MISCH, D. W. & MISCH, M. S. (1969). Reversible activation of lysosomes by dimethyl sulphoxide. *Nature, Lond.* **221,** 862–3.

PONTÉN, J., WESTERMARK, B. & HUGOSSON, R. (1969). Regulation of proliferation and movement of human glia-like cells in culture. *Exptl. Cell Res.* **58,** 393–400.

REALE, E. & LUCIANO, L. (1970). Fixierung mit Aldehyden. Ihre Eignung für histologische und histochemische Untersuchungen in der Lichtt- und Elektronenmikroskopie. *Histochemie* **23,** 144–70.

SABATINI, D. D., BENSCH, K. & BARRNETT, R. J. (1963). Cytochemistry and electron microscopy. The preservation of cellular ultrastructure and enzymatic activity by aldehyde fixation. *J. Cell Biol.* **17,** 19–58.

SMITH, R. E. & FARQUHAR, M. G. (1966). Lysosome function in the regulation of the secretory process in the cells of the anterior pituitary gland. *J. Cell Biol.* **31,** 319–47.

SCHULTZ, R. L. & KARLSSON, U. (1965). Fixation of the central nervous system for electron microscopy by aldehyde perfusion. II. Effect of osmolarity, pH of perfusate and fixative concentration. *J Ultrastruct. Res.* **12,** 187–206.

TORACK, R. M (1965) The extracellular space of rat brain following perfusion fixation with glutaraldehyde and hydroxyadipaldehyde. *Z. Zellforsch. Mikrosk. Anat.* **66,** 352–64.

TRUMP, B. F. & ERICSSON, J. L. E. (1965). The effects of the fixative solution on the ultrastructure of cells and tissues. A comparative analysis with particular attention to the proximal convoluted tubule of the rat kidney. *Lab. Invest.* **14,** 1245–323.

Cytochemical evidence for the leakage of acid phosphatase through ultrastructurally intact lysosomal membranes

ULF T. BRUNK
and JAN L. E. ERICSSON

Department of Pathology, University of Uppsala, Uppsala;
and Department of Pathology, Sabbatsberg's Hospital, Karolinska Institutet Medical School, Stockholm, Sweden

Synopsis. Fixation under 'improper' conditions of *in vitro* cultivated cells results in an extensive diffusion of the lysosomal enzyme acid phosphatase because of the influence of a low effective osmotic pressure. In the present investigation, advantage was taken of this predictable diffusion in order to establish whether or not leakage of acid phosphatase could take place through ultrastructurally 'intact' lysosomal membranes.

In order to reveal small holes in the lysosomal membranes, secondary lysosomes were labelled with thorium dioxide particles, which were presumed to appear free in the cell sap if ruptures in the membranes larger than about 100 Å were created.

The experiments revealed that following the fixation of *in vitro* cultivated human glia cells under 'improper' conditions, mitochondria and ground cytoplasm show considerable swelling artifacts, while secondary lysosomes appear to be essentially unaffected. The lysosomes, nevertheless, apparently lost most of their content of acid phosphatase, as judged from enzyme cytochemical studies. These findings indicate that leakage of acid phosphatase from ultrastructurally 'intact' lysosomes is possible.

Introduction

Since the enunciation of the lysosomal concept by de Duve, considerable interest has been focused on the possible role of lysosomes in cellular degeneration and necrosis. The lytic nature of the enzymes present in lysosomes and the occurrence of numerous compounds that could serve as substrates for these enzymes in the cytoplasm of all cells prompted the suggestion that release of lysosomal enzymes might be an early, and perhaps triggering, event in cell damage (Allison, 1968; de Duve, 1959; de Duve & Wattiaux, 1966).

Experimental evidence has accumulated to indicate that lysosomes can be 'labilized' (i.e. show an increased permeability resulting in leakage of enzymes) by a great number of substances and treatments (for review, see Weissmann, 1969). Because of the lack of suitable *in vivo* experimental systems, most of these investigations have been performed in test tubes on crude preparations of lysosomes obtained from homogenized liver. After the addition of drugs or the start of other treatments, measurements of shifts between 'free', 'sedimentable' and 'unsedimentable' activities of lysosomal enzymes were performed on the lysosome-rich fraction. In investigations of this type the leakage of hydrolytic enzymes from lysosomes during various 'stress' situations has been demonstrated (Weissmann, 1969).

The findings in these types of experiments have led to the assumption of lysosomes as potential intracellular 'suicide bags'.

The main problem in the evaluation of the validity of this theory is to establish the correct order of events leading eventually to cell death: does labilization of lysosomes and leakage of enzymes precede cell damage and contribute to degenerative processes, or is the labilization a late event occurring when the cell is already on the point of dying because of a severely disturbed metabolism?

The main support for the latter opinion is that even when cellular degeneration is fairly advanced it is in many instances still possible to demonstrate apparently intact lysosomes by means of electron microscopy. Furthermore, in sections in which the lysosomal marker enzyme acid phosphatase has been localized histochemically, granular 'staining' remains long after advanced cellular changes have become manifest. However, the possibility exists that large molecules, such as enzymes, may penetrate membranes which in the electron microscope appear intact; and a granular pattern of staining does not exclude partial leakage of enzyme from injured lyosomes.

The possibility of studying, at the ultrastructural level, a putative diffusion of lysosomal enzymes by means of enzyme cytochemistry is hampered in several ways when whole tissues are used. Problems inherent in the process of fixation, *e.g.* permeation and temperature of the fixative and artifacts created during the processing of the thin sections, often invalidate interpretation of the findings. With *in vitro* cultured monolayers of cells, these problems are diminished; penetration of the fixative is for practical purposes instantaneous, the temperature can be controlled, and no sectioning of the cells is needed before incubation.

It was found in a previous cytochemical study on the localization of acid phosphatase in *in vitro* cultured cells that considerable enzyme diffusion occurred after improper fixation (Brunk & Ericsson, 1972). In the present paper the predictable diffusion of acid phosphatase after fixation of *in vitro* cultured cells under intentionally 'improper' conditions is taken advantage of to study the relationship between the structural integrity of the lysosomal membranes on the one hand and the diffusion of this lysosomal enzyme on the other.

Material and methods

Culture conditions

All experiments were performed on *in vitro* cultured human glia cells when the cells were in early passages and showed the characteristics of Phase II according to Heyflick (1965). The cells were grown in 60 mm Falcon or Nunc plastic Petri dishes, refed twice a week with Eagle's minimal essential medium supplemented with 10% calf serum and antibiotics, and subcultivated at confluency. A detailed description of the derivation and maintenance of glia cell lines has been given previously (Pontén *et al.*, 1969).

For the experiments the cells were harvested 24–48 hr after forming a stationary monolayer at terminal density when the cells showed only minimal mitotic activity and locomotion (Abercrombie & Heaysman, 1953).

Labelling of secondary lysosomes

24 hr after forming a stationary monolayer at terminal density the cell cultures intended for labelling experiments were exposed to 0.3 ml Thorotrast*/5 ml cultivating medium for 6 hr. The cells were subsequently refed with fresh medium over-night before the start of the cytochemical experiments.

Primary fixation and washing

The cell cultures, labelled and unlabelled, were fixed *in situ* in the plastic dishes at 0°C for 60 min in either:

(1) *fixative 1*: 2% purified glutaraldehyde (Anderson, 1967; Smith & Farquhar, 1966) in 0.1 M cacodylate-HCl buffer containing 0.1 M sucrose (pH 7.2, approximately 540 mosM) (Brunk & Ericsson, 1972); or

(2) *fixative 2*: 2% purified glutaraldehyde in 0.05 M sodium cacodylate-HCl buffer (pH 7.2, approximately 340 mosM).

After fixation the cultures were washed for 60 min at 0°C in three different changes of 0.25 M sucrose—0.01 M Tris-HCl buffer (pH 7.2, approximately 270 mosM).

Enzyme cytochemistry

Acid phosphatase was demonstrated *in situ* in the plastic dishes with a Gomori-type medium (Barka & Anderson, 1962) containing 0.22 M sucrose with or without 10% dimethylsulphoxide (DMSO) added (Brunk & Ericsson, 1972). The reaction was run at 37°C, with continuous agitation of the solution, for 30, 45, 60 or 90 min (pH 5.0, osmolality approximately 1800 mosM when 10% DMSO was added and about 300 mosM when not).

The cultures intended for electron microscopy were processed as described below. For light microscopy the lead phosphate was converted to lead sulphide by treatment of the cultures for 60 sec with 1% ammonium sulphide.

In the experiments designed to test the influence on the cellular fine structure of the Gomori-type solution at 37°C and pH 5.0 the cells were incubated as described above but the lead ions were excluded from the incubation medium.

Controls for the cytochemical procedure were incubated in a complete medium containing 0.01 M NaF or in a medium lacking the substrate (β-glycerophosphate).

Electron microscopy

After glutaraldehyde fixation, or glutaraldehyde fixation followed by a cytochemical localization of acid phosphatase, the cells were post-fixed in 2% *s*-collidine-buffered osmium tetroxide, pH 7.4, for 90 min at 20°C and were subsequently prepared for electron microscopy by a method slightly modified from Biberfeld (1968). This method, which allows the study of a great number of cells with uninterrupted spatial interrelationships, has been described previously in detail (Brunk *et al.*, 1971). The cells were em-

*Thorotrast is 24% to 26% stabilized colloidal thorium dioxide by volume (size of electron dense macromolecules 10–250 Å); 25% aqueous dextrin; and 0.15% methyl parasept as preservative (Fellows Testagar, Detroit, Mich. USA).

bedded in an Araldite-Epon epoxy mixture (No. 1) according to Mollenhauer (1964), and the blocks were cut on an LKB Ultrotome or a Reichert Ultramicrotome. Thin sections mounted on formvar-coated, carbon-stabilized copper grids were studied unstained or stained with lead citrate according to Reynolds (1963) in a Siemens Elmiscope 1 A operated at 60 or 80 KV.

A summary of the experimental procedures is given in Fig. 1.

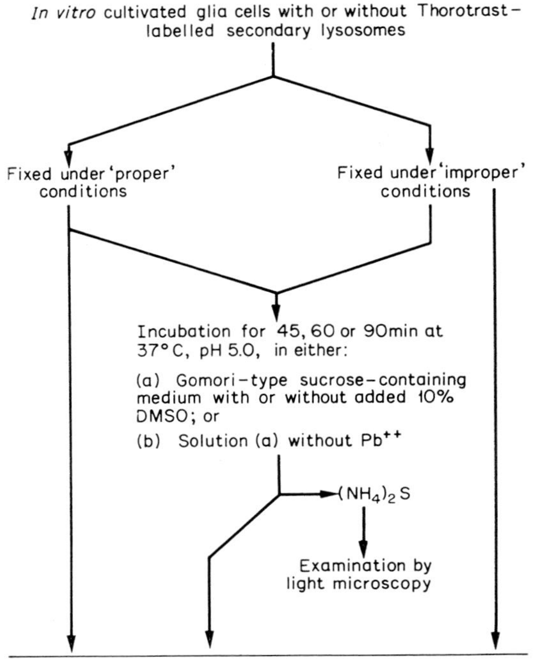

Figure 1. Summary of the experimental procedures used, as described in *Materials and Methods*.

Results

Fine structure of the glia cells

The relevant ultrastructural features of the glia cells have been described previously (Brunk *et al.*, 1971). Of special interest in the present study was the appearance of the lysosomes. These organelles were distributed diffusely within the cytoplasm, varied

Figure 2. Electron micrograph of portion of Thorotrast-labelled glia cell after fixation under 'proper' conditions. Lysosomes (L) contain densely packed electron-dense particles. Similar particles are present outside the cell in association with the microvilli (arrows), and below the cell. N = nucleus. × 24000

Figure 3. Parts of two Thorotrast-labelled glia cells after fixation under 'improper' conditions. Lysosomes appear similar to those in Fig. 2. One mitochondrion (m) shows pallor of its matrix and irregular contour; there are areas of apparent 'swelling' of the cytoplasmic ground substance (arrows), and some portions of the endoplasmic reticulum are dilated or disrupted (erd). × 24000

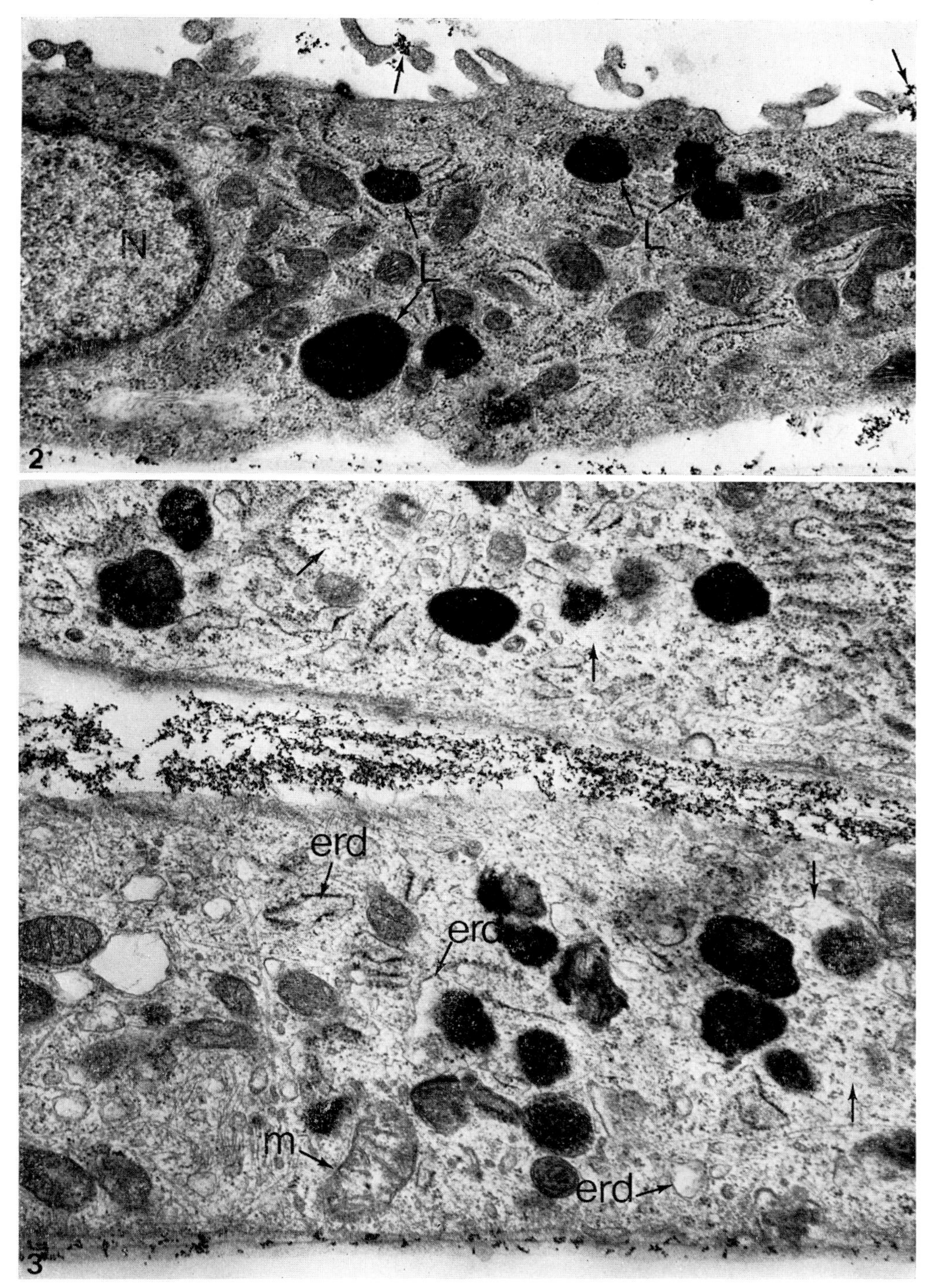
N
L
L
2
erd
erd
m
erd
3

considerably in size, and usually had an electron dense matrix in which inclusions (such as membrane fragments, granular materials and irregular densities) were suspended; they were always limited by a single, smooth-surfaced, uninterrupted membrane. Adequate preservation of the nucleus and the different organelles and elements of the cytoplasm was obtained after proper fixation (*fixative 1*). After improper fixation (*fixative 2*), irregular swelling of mitochondria, endoplasmic reticulum and cytoplasmic ground substance was usually encountered. However, the fine structure of the lysosomes was not conspicuously altered, and their limiting membranes appeared intact.

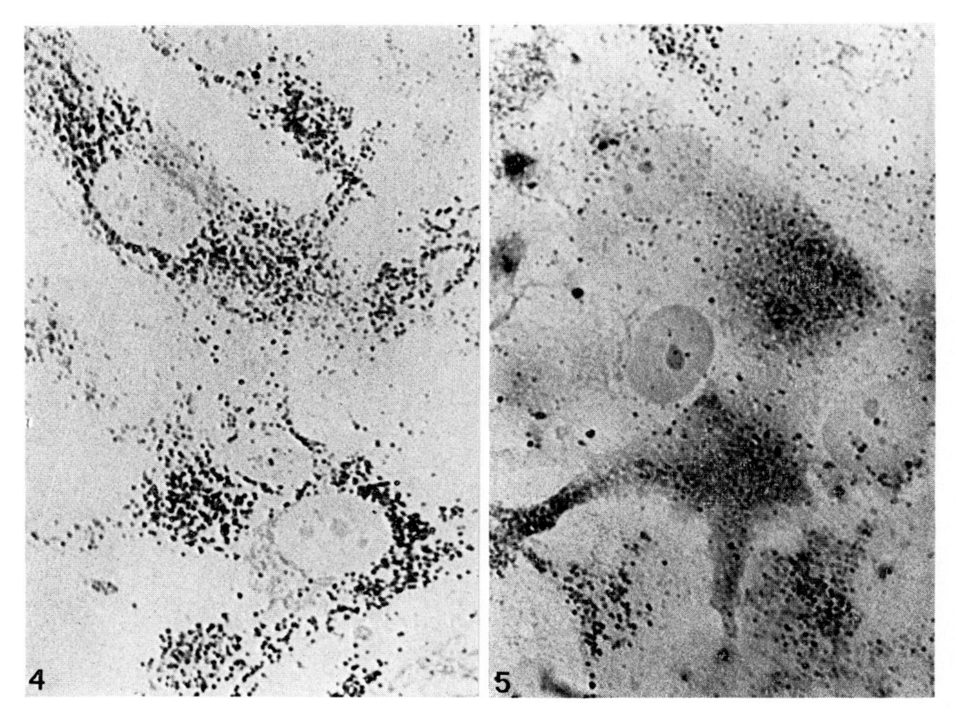

Figure 4. Light micrograph illustrating pattern of demonstrable acid phosphatase activity in the glia cells after fixation under 'proper' conditions. Incubation time, 60 min; Gomori-type medium containing 10% DMSO. Note distinct granularity in the cytoplasm and absence of diffuse nuclear or cytoplasmic 'staining'. × 750

Figure 5. Illustration of the localization of acid phosphatase in the glia cells after fixation under 'improper' conditions. Incubation as described in caption of Fig. 4. Note diffuse 'staining' of cytoplasm and nuclei, and diminished numbers of granular sites with final reaction product. × 750

Figure 6 & 7. Glia cells fixed under 'proper' conditions. Distribution of acid phosphatase visualized as described in caption to Fig. 4. Note distinct precipitates of lead phosphate in lysosomes (Fig. 6) and Golgi region (Fig. 7). Fig 6, × 16000; Fig. 7, × 32000

Figures 8–11. Glia cells fixed under 'improper' conditions. Distribution of acid phosphatase visualized as described in caption to Fig. 4. Note diffusely scattered lead phosphate precipitates in the cell sap (Figs. 8, 9 & 11) as well as lack of distinct Golgi 'staining' (Fig. 11) (G = Golgi apparatus). The lysosomes contain reaction product in variable, usually scanty, amounts (Figs. 8 & 9), although sometimes groups of lysosomes are heavily 'stained' (Fig. 10). × 24000

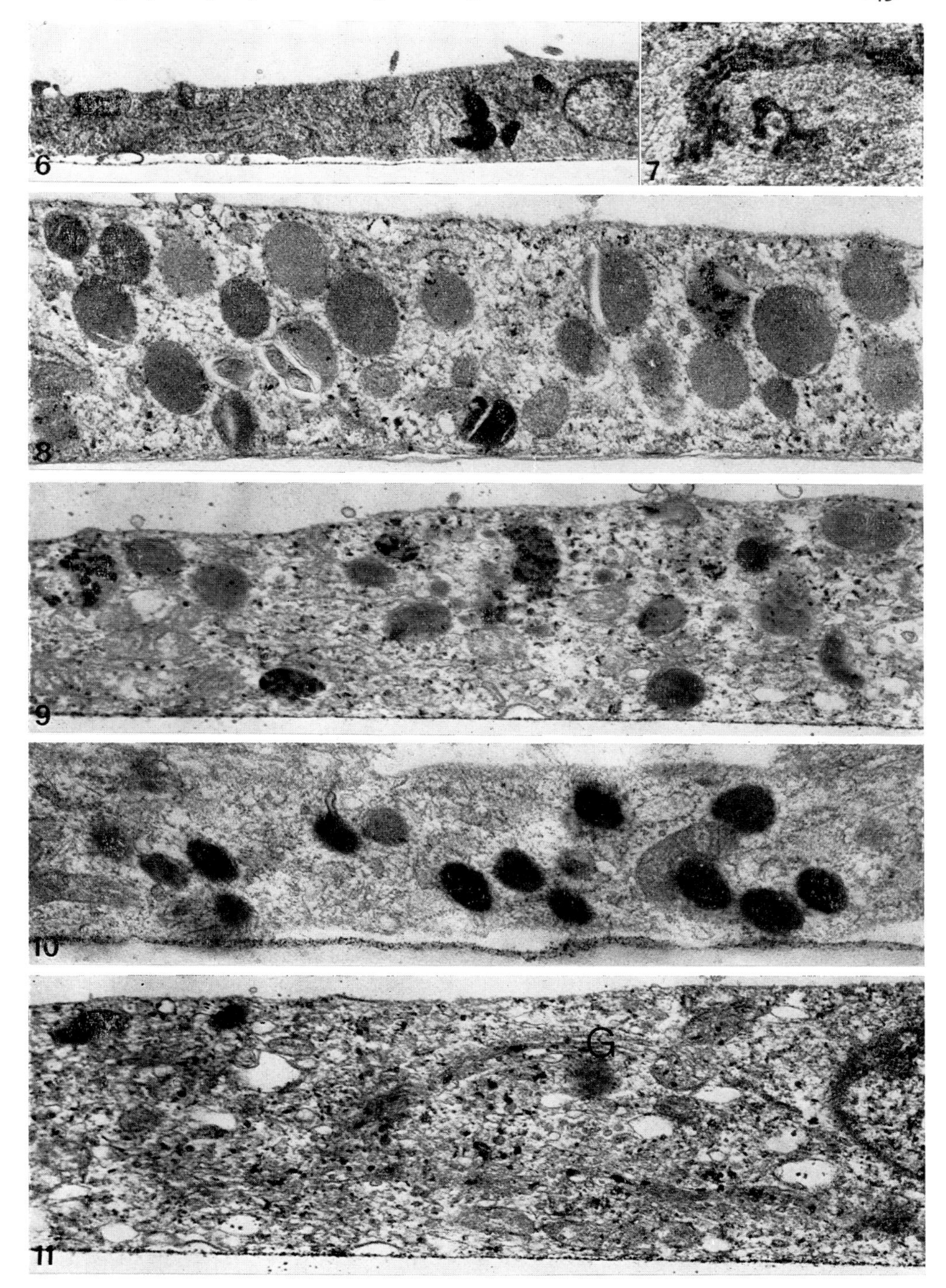
6
7
8
9
10
11
G

After exposure to thorium dioxide particles for 6 hr, virtually all secondary lysosomes contained marker particles, usually tightly packed, when the cells were fixed 16 hr after the exposure (Fig. 2). Disregarding the occurrence of thorium dioxide particles within the lysosomes, no difference in the fine structure between unexposed cells and cells exposed to marker molecules was observed. Of particular interest was the constant absence of thorium dioxide particles in the cell sap or elsewhere in the cytoplasm following both modes of fixation (Figs. 2 & 3), and the membranes limiting the secondary lysosomes were always uninterrupted.

Localization of acid phosphatase in unlabelled cells

At the light microscope level, distinct, granular deposition of final product was obtained with cells fixed with *fixative 1* (Fig. 4). When *fixative 2* was employed, diffuse cytoplasmic and nuclear 'staining' was noted in many cells (Fig. 5). In addition, granular deposits of final product occurred in some cells or in some areas of the cells (Fig. 5). Staining was absent from both properly and improperly fixed cells when they were incubated in a medium lacking β-glycerophosphate or in a complete medium containing NaF.

In the electron microscope, electron-dense lead phosphate precipitate was confined to all types of lysosomes and some of the Golgi cisternae and vesicles after using *fixative 1* (Figs. 6 & 7). The preservation of the fine structure was comparable to that in non-incubated cells; however, the contrast in the cytoplasm was markedly lower in cells that had been incubated in the Gomori medium than in those directly post-fixed in osmium tetroxide. The deposits within the lysosomes usually were diffuse and comprised tightly packed crystals covering all structures within the lysosomes.

Following improper fixation (with *fixative 2*), differing patterns of deposition of final reaction product were noted (Figs. 8–11), and some cells, or portions of cells, lacked deposits even though final reaction product was present in adjacent cells or portions of cytoplasm. When present, the deposits might be located within some or all lysosomes, and also diffusely in the cytoplasmic ground substance, without apparent connection with any of the organelles. Non-lysosomal diffuse cytoplasmic and nuclear distribution was quite common. In general the fine structure of the cells was distorted: swelling of mitochondria and ground cytoplasm and occasionally of the endoplasmic reticulum was evident. Except for complete or partial absence of final reaction product, the lysosomes were unremarkable. Hence, no ruptures of their limiting membranes were noted, nor did the matrix show evidence of dissolution or dispersion (the electron density of the matrix was similar to that in properly fixed cells).

The appearance and distribution of the final product was the same irrespective of whether or not DMSO was present in the incubation medium. However, the deposition

---

Figures 12 & 13. Thorotrast-labelled glia cells after fixation under 'improper' conditions. The cells were subsequently incubated in the modified Gomori-type medium lacking lead ions. Note signs of swelling (clear, empty areas of cytoplasmic ground substance; marked by arrows), and presence of partly disrupted mitochondria containing irregular, pale matrix areas. The lysosomes are virtually unaffected with apparently intact membranes, and contain thorium dioxide particles in variable amounts. No thorium dioxide particles are seen free in the cytoplasm. Fig. 12, × 24000; Fig. 13, × 36000

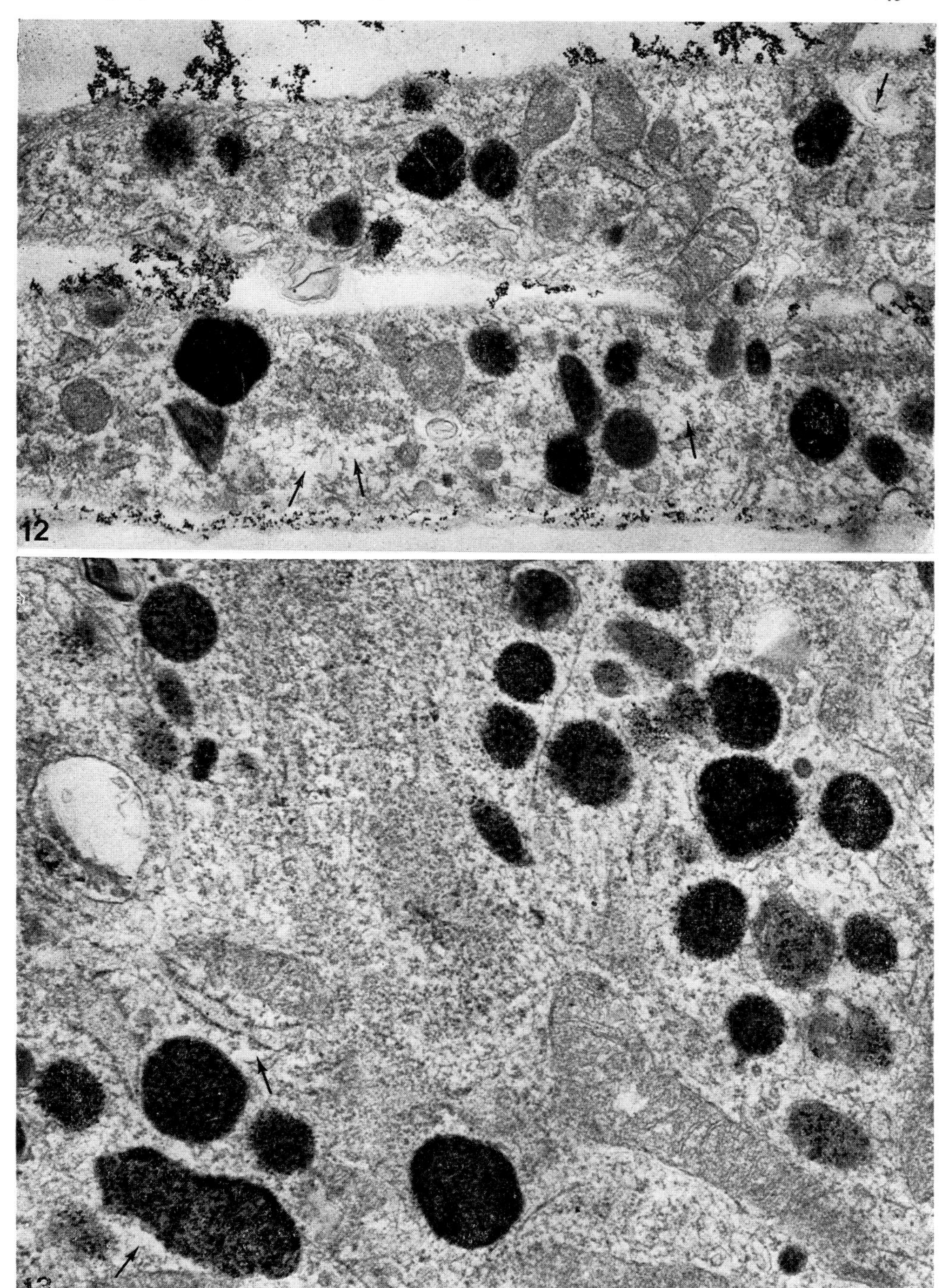
12
13

of final reaction product occurred much earlier when DMSO was utilized (Brunk & Ericsson, 1972).

Localization of acid phosphatase in Thorotrast-labelled, improperly fixed cells
Although remote, the possibility exists that improper fixation in conjunction with the incubation in the Gomori medium might result in rupture of the lyosomal membranes not revealed in the previous type of experiment. This presumed rupture might be caused by the combined effects of osmotic stress during fixation and low pH (5.0) during incubation. This possibility was tested by performing a Gomori reaction on improperly fixed cells with thorium dioxide-labelled lyosomes. However, in these experiments it became extremely difficult or impossible to discriminate between confluent thorium dioxide particles and the lead phosphate precipitate. Hence, after improper fixation it was not possible to exclude with certainty the presence of thorium dioxide particles amongst the diffusely distributed lead phosphate crystals in the cell sap. From experiments where lead ions were excluded from the incubation medium, and thus avoiding the precipitation of lead phosphate, it was clear that no thorium dioxide particles occurred in the cell sap with any of the fixation methods employed (Figs. 12 & 13).

Discussion

Lysosomes represent a group of cytoplasmic organelles limited by regular lipoprotein membranes. They contain a great variety of hydrolytic enzymes, capable of breaking down proteins, nucleic acids, complex carbohydrates and lipids (de Duve, 1959; de Duve & Wattiaux, 1966). The accumulated evidence indicates that the lysosomes are essential for the intracellular digestion both of material taken up by the cell by means of endocytosis and of cellular components during the process of autophagocytosis. It appears that this digestion results in the creation of molecules of low complexity, like amino acids and simple sugars, which may be used in the anabolic processes of the cell following their diffusion into the cell sap. Hence, it is necessary to predict the existence of 'pores' in the lysosomal membranes that allow small molecules to escape, but retain larger molecules until the degradation has been completed. This theory is in agreement with the findings in the early work of de Duve and his collaborators where they showed that the lysosomal membrane is impermeable to molecules like sucrose and glycerophosphate, while smaller molecules, such as glycerol and glucose, penetrate the membrane (Berthet *et al.*, 1951).

In comparison with glucose and glycerol, which readily penetrate the lysosomal membrane, the hydrolytic enzymes in lysosomes must be regarded as fairly large molecules; hence, there should be no risk of enzyme diffusion into the cytoplasm unless the membrane has been damaged or otherwise altered.

In studies on many types of cellular degeneration, including autolysis, it has often been suggested that leakage of hydrolytic enzymes from lysosomes occurs late in the process of cellular degeneration, and hence is not a triggering event. The reason for this assumption is that even when cellular fine structure shows considerable derangement, apparently intact lysosomes are still demonstrable. In a previous study on the effects of fixation at different osmotic pressures on the cytochemical demonstration of acid phosphatase (EC. 3.1.3.2.) it was noted that fixation at an improper osmotic pressure resulted in enzyme diffusion from the lysosomes (Brunk & Ericsson, 1972). Although it was not revealed whether this apparent diffusion was due to increased permeability of the lysosomal membranes or secondary to actual membrane disrupture, the findings did provide a model for studying the state of the lysosomal membranes after leakage of acid phosphatase into the cytoplasm.

Since small defects in the lysosomal membranes easily might escape detection, thorium dioxide particles were incorporated into the lysosomes. These particles are suitable as tracers because of their high electron density and comparatively small size (10–250 Å). Exposure of *in vitro* cultivated cells to thorium dioxide particles results in cellular uptake of the particles with subsequent segregation in secondary lysosomes (Brunk *et al.*, 1971). Considering the size of the thorium dioxide particles, they should not escape from the lysosomes unless ruptures or large holes in their membrane were present. It has been demonstrated previously that following severe damage to *in vitro* cultivated cells (induced by photosensitization injury), leakage of Thorotrast particles from the lysosomes occurs through apparent ruptures or holes in the lysosomal membranes (Ericsson & Brunk, 1972). Hence, the particles are not firmly bound to the lysosomal matrix. The presence of thorium dioxide particles is easily demonstrated in the cell sap following such lysosomal injury (Ericsson & Brunk, 1972).

In the present study, no leakage of thorium dioxide particles was noted following improper fixation at a low effective osmotic pressure. Furthermore, the lysosomes showed an appearance similar to that of properly fixed cells, while other components of the cell, such as the mitochondria and the ground cytoplasm, were less well preserved after improper fixation than after optimum fixation conditions. Thus, morphologically the lysosomes appeared to be more resistant to unfavourable osmotic conditions than other elements in the cytoplasm. Nevertheless, the cytochemical evidence clearly indicated that there was a large-scale escape of acid phosphatase from the lysosomes. It appears most likely that this escape occurred in conjunction with the improper fixation procedure. However, the findings do not exclude with absolute certainty the possibility that acid phosphatase leaked out of the lysosomes during the incubation in the substrate medium (because of the combined effect of hypotonic stress during fixation and low pH during incubation). However, this seems all the more unlikely considering the fact that the enzymes must have been affected by the fixative, and probably immobilized.

Judging from the findings in those experiments where the cells were incubated in a modified Gomori-type medium lacking lead ions, the escape of acid phosphatase from the lysosomes was not due to ruptures in their membranes created at the time of incubation. Hence, the observations in these experiments show that one of the lysosomal enzymes, acid phosphatase, can disappear from morphologically intact lysosomes, probably as a result of hypo-osmotic damage. It is logical to assume, provided the other lysosomal enzymes are not bound in a different way in the lyosomes, that these enzymes also can escape to the cell sap through uninterrupted membranes.

As was mentioned previously, it has often been claimed that leakage of hydrolytic enzymes from injured lysosomes is probably a late event in cellular degeneration, because it is possible to demonstrate virtually intact lysosomes containing cytochemically demonstrable acid phosphatase or other lysosomal marker enzymes even when cellular degeneration is pronounced. However, on the basis of such findings it is not possible to rule out the existence of a permeability change that allows a certain escape of enzyme from the lyosomes. If a slight or moderate release occurs, it is not necessarily revealed by cytochemical studies of lysosomal enzymes in fixed tissues, since only about 10% of the initial activity of acid phosphatase remains after proper fixation (Arborgh *et al.*, 1971; Brunk & Ericsson, 1972). In the present investigation, extensive leakage of acid phosphatase from the lysosomes appeared to have taken place, since many lysosomes were devoid of reaction product and there was diffuse staining of the cytoplasm in many cells following improper fixation. It is possible that partial loss of enzyme from the lysosomes is reflected cytochemically by the occurrence of areas free of final reaction product.

Although leakage of acid phosphatase through intact membranes seems to be proven in the present system, it still remains to be clarified if, similarly, escape of hydrolytic lysosomal enzymes can occur in living cells as a result of other types of injury, such as anoxia, ischemia, acidosis or drug action. It is evident, however, that the preserved fine structure of the lysosomes, or even the discrete cytochemical localization of acid phosphatase within the lysosomes still remaining, does not exclude fairly extensive loss of enzyme.

Acknowledgements

The skilled assistance of Miss Silwa Mengarelli, Mrs Britt-Marie Åkerman, Mr Bengt-Arne Fredriksson, and Mr Magnus Norman is gratefully acknowledged.

This work was supported by grants from the Swedish Cancer Society (grant number 71:208) and the Swedish Medical Research Council (projects B71-12X-1006-06C and B72-12X-1007-07A).

References

ABERCROMBIE, M. & HEAYSMAN, J. E. M. (1953). Observations on the social behaviour of cells in tissue culture. I. Speed of movement of chick heart fibroblasts in relation to their mutual contacts. *Exptl. Cell Res.* **5**, 111–31.

ALLISON, A. C. (1968). Lysosomes. In: *The Biological Basis of Medicine* (eds. E. E. Bittar and N. Bittar), Vol. 1, p. 209. Academic Press: London and New York.

ANDERSON, P. J. (1967). Purification and quantitation of glutaraldehyde and its effect on several enzyme activities in skeletal muscle. *J. Histochem. Cytochem.* **15**, 652–61.

ARBORGH, B., ERICSSON, J. L. E. & HELMINEN, H. (1971). Inhibition of renal acid phosphatase and aryl sulfatase activity by glutaraldehyde fixation. *J. Histochem. Cytochem.* **19**, 449–51.

BARKA, T. & ANDERSON, P. J. (1962). Histochemical methods for acid phosphatase using hexazonium pararosanilin as a coupler. *J. Histochem. Cytochem.* **10**, 741–53.

BERTHET, J., BERTHET, L., APPELMANS, F. & DE DUVE, C. (1951). Tissue fractionation studies. 2. The nature of the linkage between acid phosphatase and mitochondria in rat-liver tissue. *Biochem. J.* **50**, 182–9.

BIBERFELD, P. (1968). A method for the study of monolayer cultures with preserved cell orientation and interrelationship. *J. Ultrastr. Res.* **25**, 158–9.

BRUNK, U., ERICSSON, J. L. E., PONTÉN, J., & WESTERMARK, B. (1971). Specialization of cell surfaces in contact-inhibited human glia-like cells *in vitro*. *Exptl. Cell Res.* **67**, 407–15.

BRUNK, U. & ERICSSON, J. L. E. (1972). Demonstration of acid phosphatase in *in vitro* cultured cells. Significance of fixation, tonicity and permeability factors. *Histochem. J.* **4**, 349–63.

DE DUVE, C. (1959). Lysosomes, a new group of cytoplasmic particles. In: *Subcellular Particles*, p. 128. (ed. T. Hayashi). New York: The Ronald Press Co.

DE DUVE, C. & WATTIAUX, R. (1966). Functions of lysosomes. *Ann. Rev. Physiol.* **28**, 435–92.

ERICSSON, J. L. E. & BRUNK, U. (1972). Alterations in lysosomal membranes as related to diseases processes. In: *Membrane Pathology* (eds. B. F. Trump and A. Arstila). New York: Academic Press. In press.

HAYFLICK, L. (1965). The limited *in vitro* lifetime of human diploid cell strains. *Exptl. Cell Res.* **37**, 614–36.

MOLLENHAUER, H. H. (1964). Plastic embedding mixtures for use in electron microscopy. *Stain Technol.* **39**, 111–14.

PONTÉN, J., WESTERMARK, B. & HUGOSSON, R. (1969). Regulation of proliferation and movement of human glia-like cells in culture. *Exptl. Cell Res.* **58**, 393–400.

REYNOLDS, E. S. (1963). The use of lead citrate at high pH as an electron-opaque stain in electron microscopy. *J. Cell Biol.* **17**, 208–12.

SMITH, R. E. & FARQUHAR, M. G. (1966). Lysosome function in the regulation of the secretory process in the cells of the anterior pituitary gland. *J. Cell Biol.* **31**, 319–47.

WEISSMANN, G. (1969). The effects of steroids and drugs on lysosomes. In: *Lysosomes in Biology and Pathology*. (eds. J. T. Dingle and H. S. Fell), Vol. 1, p. 276. Amsterdam, London: North-Holland.

Improved preservation of alkaline phosphatase in salivary glands of the cat

K. J. DAVIES *and* J. R. GARRETT

Department of Oral Pathology,
The Dental School,
King's College Hospital Medical School,
London

Synopsis. The effects of tissue preparation on the localization of alkaline phosphatase were assessed at the light and electron microscopical levels. Glutaraldehyde-containing solutions were more inhibitory than formaldehyde alone but appeared to give better *in situ* immobilization of the enzyme. The inhibitory effects of glutaraldehyde solutions were related especially to temperature and time and not necessarily to material absorbing at 235 nm. Distilled glutaraldehyde was the only form of glutaraldehyde to give consistently good results with least inhibition. Snap-freezing of pre-fixed tissue, after washing in a sucrose-containing solution, gave satisfactory results without undue ice-crystal formation. Immersion in dimethylsulphoxide before snap-freezing gave less good localization with greater diffusion of reaction product. Use of a Sorvall tissue-chopper on unfrozen tissue did not produce satisfactory sections. Free-floating sections of pre-fixed material appeared less inhibited than glass-mounted sections. The most satisfactory results were obtained after per-arterial perfusion with a 1% distilled glutaraldehyde-0.8% formaldehyde mixture, containing dextran, for up to 5 min followed by formaldehyde-dextran. The results were uniform; there was stronger staining of stromal surfaces of myoepithelial cells than previously, basal duct cells were also stained and occasional staining between acinar cells was now evident for the first time.

Introduction

In a previous study (Garrett & Harrison, 1970) it was demonstrated that alkaline phosphatase was present on the plasma membrane of salivary myoepithelial cells in the cat. However, variations in the end results sometimes appeared to be related to the method of tissue preservation. Thus, occasional basal duct cells also showed a positive reaction in 'post-fixed' cryostat sections whereas they were not detected in 'pre-fixed' material.

The purpose of the present investigation has been to study the effects of different methods of fixation, tissue storage and section cutting, on the localization of alkaline phosphatase at the light and electron microscopical levels with the aim of finding the best procedure to give the most reproducible results with good localization and minimal inhibition.

In the previous work (Garrett & Harrison, 1970), the glutaraldehyde solutions were made up from TAAB E.M. grade glutaraldehyde and at the time this seemed satisfactory but at the outset of the present investigation a new bottle of the same grade glutaraldehyde appeared to be totally inhibitory and so we were impelled to include in the present study an investigation of the effects of impurities in the glutaraldehyde on the results.

Materials and methods

Tissues

Tissue was removed from submandibular, sublingual and parotid glands of mature cats under Nembutal anaesthesia. As far as was practical, as many of the preservation procedures listed below were tested on the same tissue for comparative purposes.

Fixatives

Formaldehyde, glutaraldehyde and mixtures of the two were tested. Formaldehyde was always made up freshly from paraformaldehyde. Glutaraldehyde was used as supplied (E.M. grade; TAAB) or purified by charcoal washing or distillation (see below). All fixatives were made up in cacodylate buffer, pH 7.2, with final strength 0.08 M and containing 3 mM $CaCl_2$.

Formaldehyde alone was made up to 4% and usually contained 7.5% sucrose. Occasionally, 2 or 4% formaldehyde containing 2–4% dextran (average mol. wt. 70 000) was used.

5% glutaraldehyde was sometimes used initially but tissue morphology on examination by electron microscopy was invariably poor and mixtures of formaldehyde and glutaraldehyde gave much better morphology and less enzyme inhibition. Subsequently, therefore, glutaraldehyde was always used in mixtures with formaldehyde and made up as described by Karnovsky (1965) giving 4% formaldehyde, 5% glutaraldehyde, or in half or one-fifth this strength. Sometimes 2–4% dextran (average mol. wt. 70 000) was added to these mixtures.

Purification of glutaraldehyde

By u.v. spectrophotometry it was confirmed that there was massive absorption at 235 nm in all the commercial samples tested. The glutaraldehyde was 'purified' by washing with Norit-Ex charcoal or by vacuum distillation according to the methods described by Anderson (1967). Concentrations of the glutaraldehyde solutions resulting therefrom were assayed by the method of Frigerio & Shaw (1969) before making up the final fixative solution. They were also tested by u.v. spectrophotometry to confirm an absence of absorption at 235 nm.

Methods of fixation

'Post-fixation' of sections of frozen, unfixed tissue cut on a cryostat and mounted on glass slides was used for light microscopy or 'pre-fixation' by immersion or per-arterial perfusion was used before sectioning for light or electron microscopy.

Immersion fixation was performed on small blocks with sides of less than 3 mm for periods of 1–20 hr at 4°C or room temperature.

Perfusion fixation was usually performed bilaterally and the simplest and most effective technique was through a cannula inserted in the external carotid artery on the cardiac side of the lingual artery. Perfusion fluids were held in bottles at a height of about 5 ft above the animal. At the start of the perfusion, oxygenated Ringer-Locke solution at 37°C was washed through (10–20 sec) followed by the fixative fluid at room temperature. The abdominal aorta and inferior vena cava were opened to exsanguinate the animal at the commencement of the perfusion. After a short interval the external jugular veins were cut and, after clear fluid had passed through the abdominal vessels they were clamped to localize the perfusion to the glandular region. Perfusion of one fixative was sometimes followed by a second fixative. Perfusion was usually continued for 10 min and then the glands were rapidly removed, cut into small pieces and placed into similar fixative for up to 30 min or sometimes longer before subsequent treatment. It should be noted that only purified fixatives were used for perfusion fixation.

Washing of pre-fixed tissue

This was performed at 4°C with 0.05 M cacodylate buffer, pH 7.2, containing 7.5% sucrose for 1–18 hr. At least two changes of the washing solution were made. Sometimes the washing solution also contained 10% dimethyl sulphoxide (DMSO).

Snap-freezing of the tissues

Blocks of unfixed tissue or pre-fixed tissues removed from the washing solution were rapidly frozen in *iso*-pentane cooled by liquid nitrogen and then stored in a −70°C refrigerator until subsequent use.

Section cutting

Cryostat sections were cut at 10 μm of frozen unfixed or pre-fixed tissue and mounted on glass slides for light microscopical study. Sections of frozen pre-fixed material were cut on a freezing microtome at 50 μm and handled as free-floating sections for electron microscopical study with some sections being examined as a light microscopical control.

Sections of unfrozen pre-fixed material were cut on a Sorvall tissue-chopper set at 50–100 μm and handled as free-floating sections for electron microscopical study.

Histochemical procedures

The azo-dye technique of Stutte (1967) using naphthol-AS-BI phosphate and hexazotized New Fuchsin was employed at pH 9.2 for light microscopical studies. The direct lead method of Hugon & Borgers (1966*a*,*b*) with β-glycerophosphate at pH 9 was used for light and electron microscopical studies. Incubations of up to 20 min were used at room temperature or sometimes at 4°C. For electron microscopy, 7.5% sucrose was included in the media.

Sections for light microscopy were mounted in DPX after the azo-dye technique and in colophonium after the lead technique. Sections for electron microscopy were post-fixed in 1% osmium tetroxide for 30 min at 4°C and then dehydrated in alcohols and embedded in Araldite. Ultra-thin sections were examined unstained in an AEI EM6B electron microscope operated at 60 kV.

Results

The results have been assessed subjectively in respect to the intensity, localization and uniformity of staining by the enzyme reaction product in each section examined.

LIGHT MICROSCOPY (Figs. 1 & 2)

(1) *Post-fixation*

The strongest staining for alkaline phosphatase activity was seen in cryostat sections post-fixed in formaldehyde-sucrose, but the staining was not always precise. Myoepithelial cells were strongly stained and were seen embracing acini and intercalary ducts in sections of all glands. Variations occurred in the intensity of myoepithelial cell

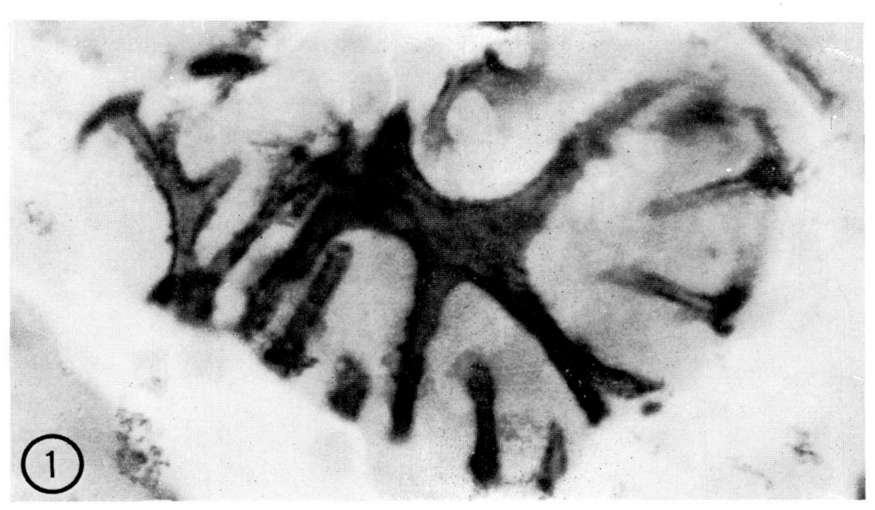

Figure 1. Sublingual gland, stained by azo-dye method. Section 'post-fixed' by Karnovsky's fixative. Strong staining of myoepithelial cell and processes embracing an acinus is evident. × 960

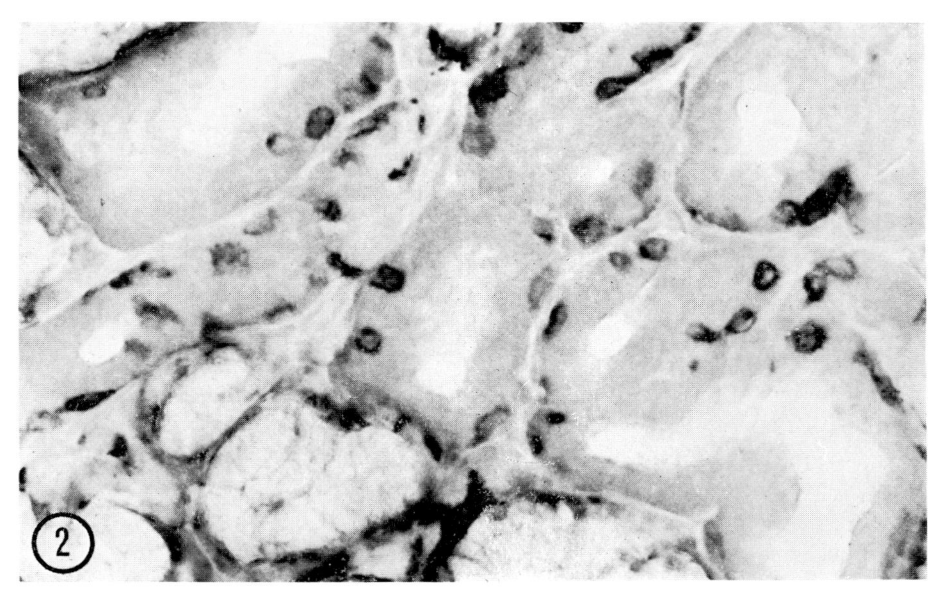

Figure 2. Submandibular gland stained by azo-dye method. Section 'post-fixed' by formaldehyde-sucrose. Variable staining of basal duct cells is evident. × 450

staining in any one section. Some blood vessels were also stained in each section. Basal duct cells, in striated ducts and interlobular ducts, were stained and thereby became clearly visible in submandibular and parotid tissues. They were irregular in their distribution, varied in number from one animal to another and were more numerous in the submandibular gland, presumably due to the greater number of striated ducts in this gland. There was great variation in the intensity of staining in these cells and the staining was rarely as strong as in the myoepithelial cells.

Fixation with glutaraldehyde-containing solutions was always more inhibitory than formaldehyde alone in respect to the amount of reaction product that formed. However, the localization of reaction product was often more precise after fixation in glutaraldehyde-containing solutions. The inhibitory effects were related especially to the temperature and time of fixation and also to concentration. Grossly impure glutaraldehyde tended to be the most inhibitory but the results were variable and did not always reflect the amount of material absorbing at 235 nm. Distilled glutaraldehyde gave the most consistently good results. Charcoal-washed glutaraldehyde gave somewhat variable results despite an absence of material absorbing at 235 nm. Glutaraldehyde-fixed sections often showed a patchy staining. The best results with a glutaraldehyde-containing solution were obtained with one-fifth strength Karnovsky's (1965) mixture applied for 1 min at 4°C, made up from distilled glutaraldehyde and containing 2% dextran. The staining was uniform and precise but was usually less strong than 5 min fixation in 4% formaldehyde sucrose at 4°C.

Sometimes, kidney sections were studied at the same time but the inhibitory effects of glutaraldehyde were never so conspicuous as in the salivary tissue. This seemed to relate to the greater activity of alkaline phosphatase present in the renal tissues.

The azo-dye technique and the lead technique gave similar results but the azo-dye method seemed more sensitive and sometimes appeared to stain more basal cells. It also tended to give more precise staining in the best sections.

It should be mentioned that even in the most uniformly stained sections variations in the intensity of myoepithelial cell staining was always evident.

(2) *Pre-fixation*

In glass-mounted sections of 'pre-fixed' material the inhibitory effects were usually more pronounced than in 'post-fixed' sections, especially after immersion fixation. The inhibitory pattern was similar to that for 'post-fixation' but the differences between the least inhibited tissues after formaldehyde-sucrose or formaldehyde-dextran and glutaraldehyde-containing fixatives was inconspicuous in respect to the intensity of staining. The use of distilled glutaraldehyde gave the most consistently good results with glutaraldehyde-containing solutions. Diffusion of staining was less obvious with formaldehyde 'pre-fixation' than with 'post-fixation'.

Zoning of activity was often seen after immersion 'pre-fixation' with glutaraldehyde-containing solutions, especially in larger blocks, with most staining in the intermediate zone. This feature was most conspicuous in the parotid gland and least evident in the sublingual gland. Zoning was always more evident with impure glutaraldehyde and this may have had a closer relationship with the amount of material absorbing at 235 nm than the more general inhibitory effect seemed to have. Glutaraldehyde alone caused more zoning and more inhibition than glutaraldehyde-formaldehyde mixtures. Perfusion 'pre-fixation' gave more uniform results.

Glass-mounted cryostat sections of 'pre-fixed' material sometimes showed a peculiar irregularity of staining with small patches of inhibition that varied in position in serial sections of the same block. This was always more evident in glutaraldehyde-fixed sections. However, free-floating 50 μm sections of the same material, used as light microscopical controls for electron microscopical preparations were always more uniformly stained. Furthermore, they often showed moderately strong staining even if a glass-mounted section of the same material appeared to be almost totally inhibited. From this it is deduced that drying the 'pre-fixed' sections on the slide increases the inhibition of alkaline phosphatase activity in an irregular manner especially in glutaraldehyde pre-fixed material.

Tissue from blocks frozen from washing solutions containing DMSO always showed more diffuse staining than blocks of the same tissue in which DMSO had not been used. Sectioning of the material containing DMSO presented considerable difficulty in obtaining good sections in a conventional cryostat.

The best results after immersion pre-fixation were obtained with low concentration glutaraldehyde-formaldehyde mixtures containing dextran, e.g. one-fifth strength Karnovsky's fixative made up from distilled glutaraldehyde. Formaldehyde-dextran seemed to give marginally better results for immersion fixation than formaldehyde-sucrose. Perfusion fixation with similar mixtures gave the best results light microscopically after pre-fixation but will be considered in more detail in the section on electron microscopy.

As with 'post-fixed' tissue, variations in the intensity of myoepithelial staining were present after 'pre-fixation', even in the most uniformly stained sections showing the least inhibition. Basal duct staining was, at best, faint and often non-existent, especially after more inhibitory mixtures.

ELECTRON MICROSCOPY (Figs. 3–8)

(1) *Immersion fixation*

After immersion fixation the results were comparable to those seen light microscopically and the inhibitory pattern was similar. However, a little activity could usually be detected even if cryostat sections appeared totally inhibited. Distilled glutaraldehyde was the only reliable glutaraldehyde to give moderately consistent results.

At best, there was staining on the outer surface of the plasma membrane of the myoepithelial cells, especially those adjacent to underlying parenchymal cells. Vesicles in the myoepithelial cell, close to both parenchymal and stromal surfaces, also tended to be stained. Otherwise the stromal surface was weakly stained, if at all. A diffuse precipitate was sometimes associated with the basement membrane nearby and sometimes extended on to adjacent structures. Such staining is the sort referred to as 'diffuse staining' in future reference. Some capillaries showed reaction product in vesicles within their endothelial cells.

The tissue morphology in sections cut on a freezing microtome from pre-fixed blocks snap-frozen after washing in a sucrose-containing solution were generally good and ice crystal artifact was not a conspicuous feature. Washing in DMSO before freezing or cutting sections of unfrozen material on a Sorvall tissue chopper did not cause any marked improvement in overall morphology.

Localization of reaction product after immersion in DMSO tended to be much more diffuse, especially in basement membranes and interstitial regions. Great difficulty was experienced in cutting uniform thin sections with the Sorvall tissue-chopper, possibly because of the lobular nature of the tissue being handled. As a consequence, uneven staining resulted, some parts being good and other parts being poor, with no staining at all, and often diffuse deposition of reaction product was evident in the interstitial tissues.

(2) *Perfusion fixation*

Perfusion fixation undoubtedly gave the best results at the electron microscopical level whenever the perfusion had been satisfactory. The results were more uniform, with better preservation of tissue morphology and better localization of reaction product than when corresponding solutions had been used for immersion fixation.

Dextran was added to the fixative solutions in all the perfusions, as suggested by Bohman & Maunsbach (1970), and only purified glutaraldehyde was used in the glutaraldehyde-containing solutions.

Frozen sections gave satisfactory results. Tissue washed in DMSO mixtures gave more diffuse staining than those not so treated. Results with the Sorvall tissue chopper were not very satisfactory for the reasons given in the previous section.

(a) 4% *Formaldehyde-dextran* (Fig. 3). The results were better than from tissues immersion-fixed in formaldehyde solutions. Somewhat stronger staining tended to occur on the stromal plasma membrane of myoepithelial cells than was evident after any of the immersion fixations. In some areas the staining was reasonably precise but in other parts

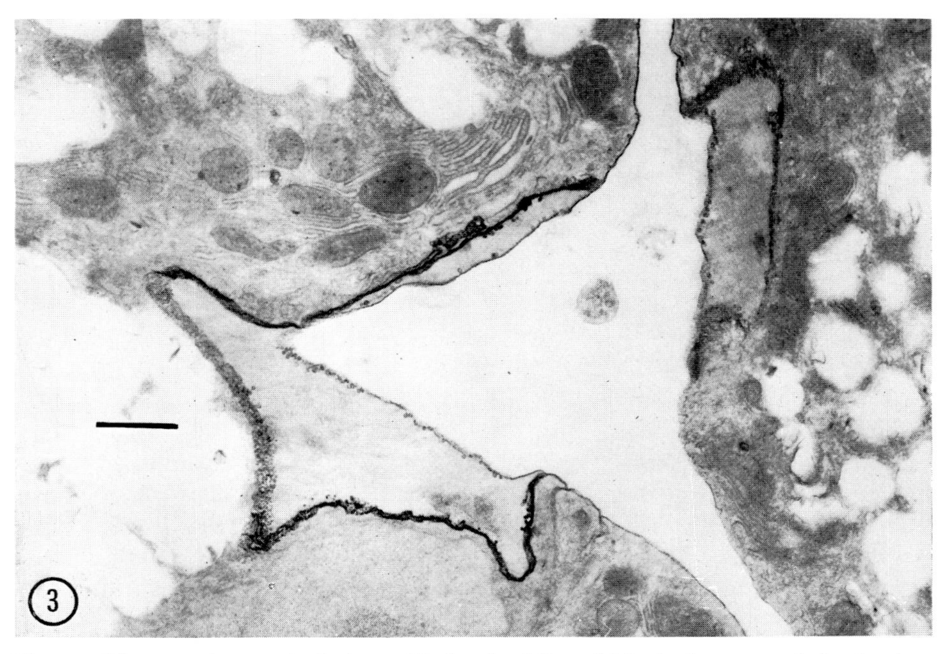

Figure 3. Electron micrograph of submandibular gland (formaldehyde-dextran perfusion fixation; frozen section) showing reaction product round the plasma membrane of two myoepithelial cells; one cell shows more activity than the other. N.B. Other parts of the same section showed much more diffusion. Bar = 1 μm.

there was much diffusion of reaction product. Some capillaries showed staining and occasionally an irregular staining was found between adjacent basal folds of striated ducts but no stained basal duct cells have been detected. There was often great variation in the amount of myoepithelial cell staining in adjacent myoepithelial cells.

(b) *Glutaraldehyde-formaldehyde-dextran mixtures* (Figs. 4–8). 10 min perfusion with half-strength Karnovsky's fixative, containing dextran, caused much more inhibition of alkaline phosphatase activity than formaldehyde-dextran but the overall morphological preservation was better. 10 min with fifth-strength Karnovsky's fixative, containing dextran, caused somewhat less inhibition and the localization of enzyme reaction product

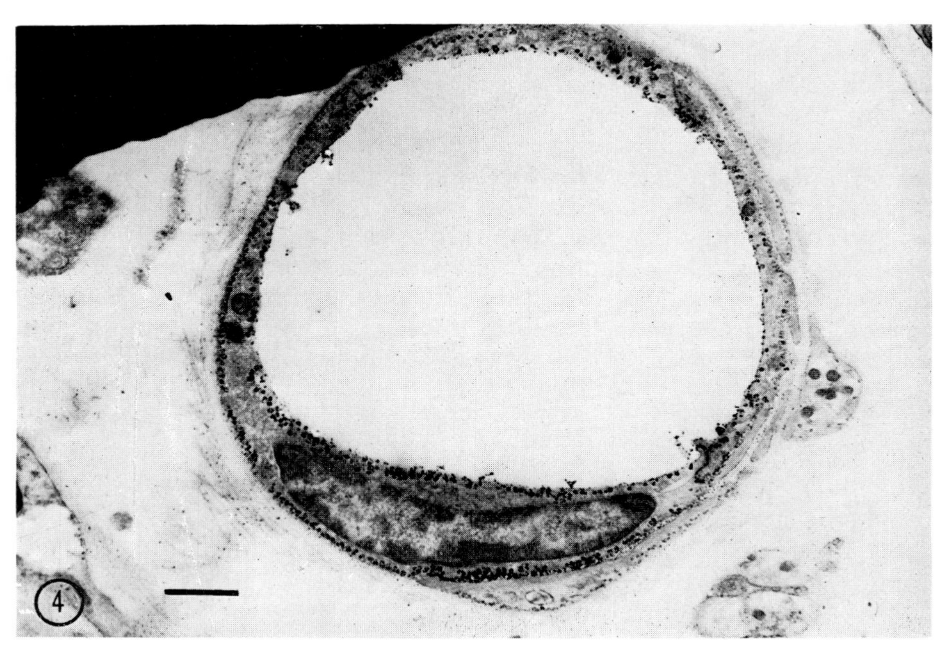

Figure 4. Electron micrograph of submandibular gland (perfusion fixation one-fifth Karnovskys' fixative plus dextran followed by 2% formaldehyde-dextran, i.e. 'double-perfusion'; frozen section) showing final reaction product in vesicles in endothelial cells of a capillary and between cellular junctions. Bar = 1 μm.

seemed to be improved. It was, therefore, decided to try a low concentration glutaraldehyde formaldehyde mixture containing dextran for a short while and continue the perfusion with a formaldehyde-dextran solution, with the aim of washing out any surplus glutaraldehyde before it could cause too much inhibition yet allowing milder formaldehyde fixation to continue. This 'double perfusion' has achieved the best results of all and the most satisfactory combination to date has been one-fifth strength Karnovsky's fixative (made up with distilled glutaraldehyde) plus 4% dextran for 5 min followed by 2% formaldehyde-4% dextran for 5 min and subsequent immersion in the same for up to 30 min. After this fixation both tissue morphology and localization of the enzyme reaction product were good and uniformly reproducible. Stronger staining was seen on

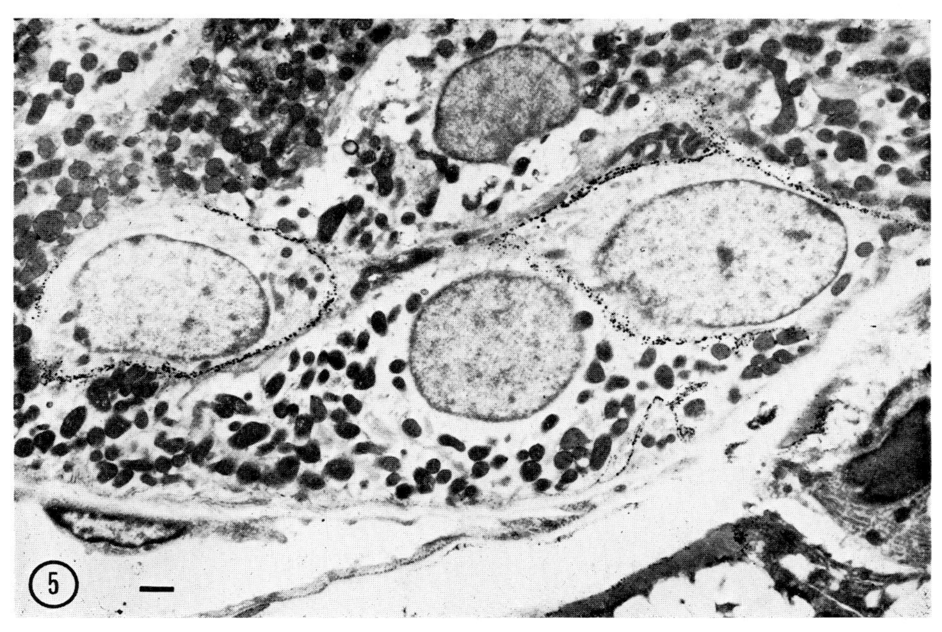

Figure 5. Electron micrograph of submandibular gland ('double perfusion' fixation; tissue-chopper section) showing final reaction product round basal duct cells in a striated duct. Note the weak staining of the myoepithelial cell in the bottom of the picture. Bar = 1 μm.

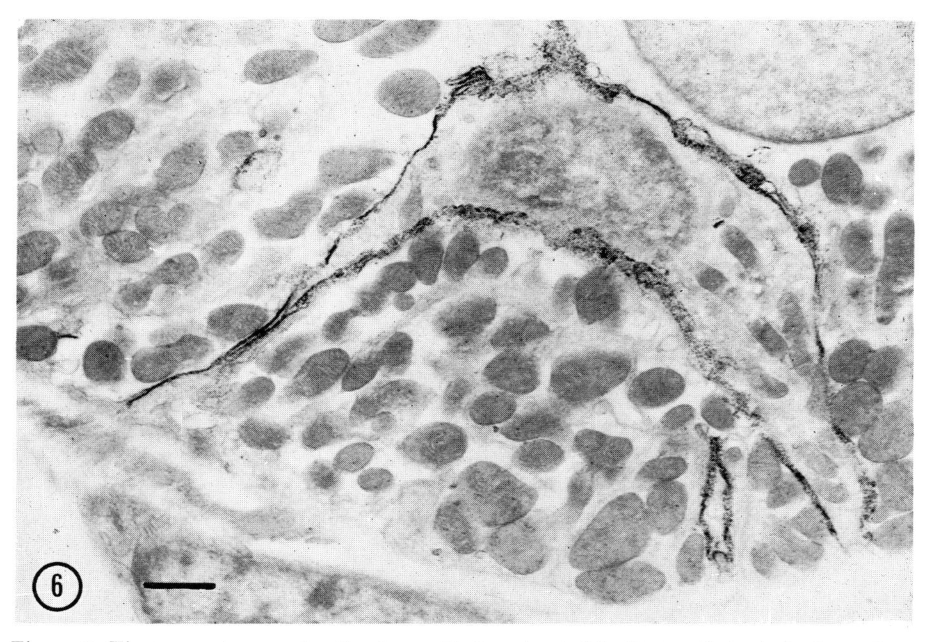

Figure 6. Electron micrograph of submandibular gland ('double perfusion' fixation; frozen section) showing final reaction product round a basal duct cell (cut tangentially) in a striated duct. Bar = 1 μm.

the stromal surface of the plasma membrane of myoepithelial cells than previously. Variations still occurred in the amount of reaction product around adjacent myoepithelial cells (see Fig. 7). Reactive capillaries showed well localized activity in vesicles in their endothelial cells and between adjacent endothelial cells (Fig. 4). Basal duct cells now showed activity round their parenchymal plasma membrane (Figs. 5, 6) but little has been detected on the stromal surfaces. Finally, in some situations (Figs. 7, 8) activity

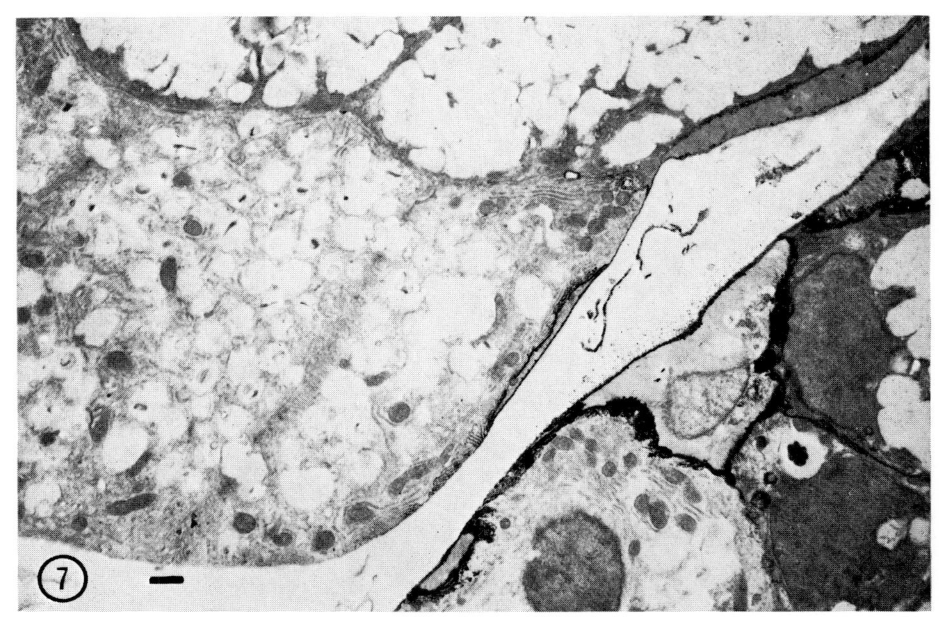

Figure 7. Electron micrograph of submandibular gland ('double perfusion' fixation; frozen section) showing variable amounts of final reaction product round myoepithelial cells, some of which show moderate amounts on their stromal surfaces. Deposit is also present between two acinar cells on the right. Bar = 1 μm.

has been clearly seen between contiguous surfaces of acinar cells and on their luminal surfaces, especially in the sublingual gland. No indication of the possibility of this localization had been given by any of the previous methods.

General comment

Throughout this work, irrespective of the methods of fixation, tissue preservation, incubation or examination, it has been evident that the staining of myoepithelial cells for alkaline phosphatase activity has tended to vary between the different glands of the same animal. Staining has usually been most readily performed with strongest reaction and least susceptibility to inhibition in the sublingual gland. In the parotid gland the myoepithelial cells have often shown the weakest staining and the greatest susceptibility to inhibition. Those in the submandibular gland have usually been somewhere between these two extremes in their reactivity although sometimes they have been similar to those of the sublingual glands.

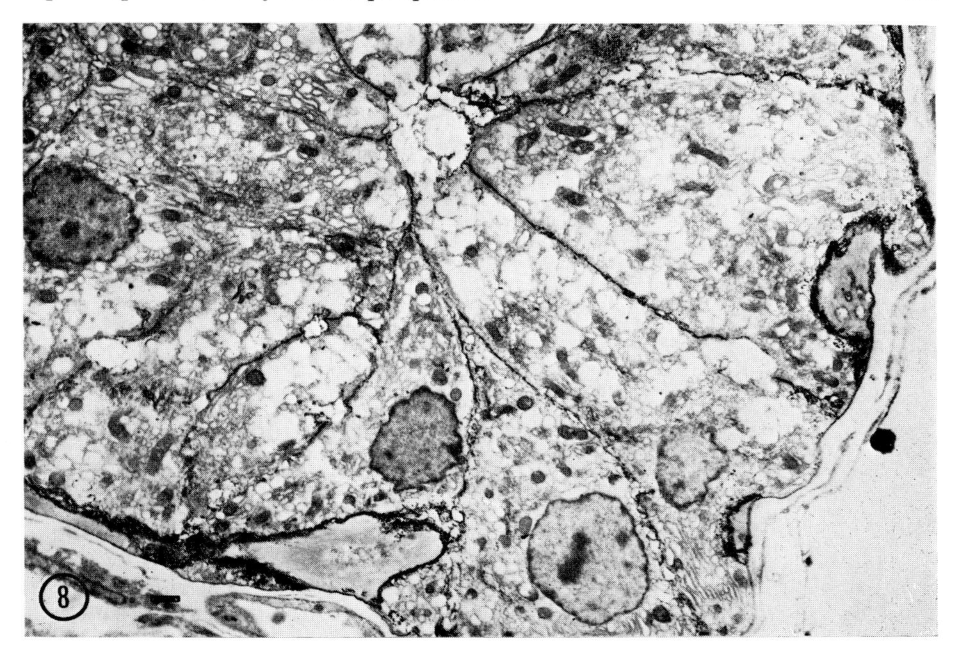

Figure 8. Electron micrograph of sublingual gland ('double perfusion' fixation; frozen section) showing an acinus with final reaction product round peripherally-placed myoepithelial cells, between acinar cells and on the luminal surfaces of acinar cells. Bar = 1 μm.

Discussion

In general, fixation with formaldehyde caused less inhibition of alkaline phosphatase in salivary tissue for cytochemical examination than glutaraldehyde; mixtures of the two tended to give intermediate results. The latter finding may relate to the observation of Hopwood (1970) that, with bovine serum albumin, glutaraldehyde induced the greatest formation of polymers but with formaldehyde it mainly remained as monomer and formaldehyde-glutaraldehyde mixtures gave intermediate results. This was interpreted as evidence of competition between the two aldehydes for the same amino-acid residues on the albumin. Similar competition may allow formaldehyde to prevent some configurational alterations of alkaline phosphatase by glutaraldehyde during fixation in solutions containing mixtures of the two, thereby protecting some of its enzyme activity.

The only reproducibly satisfactory glutaraldehyde was that made from freshly distilled glutaraldehyde. Charcoal-washed samples of glutaraldehyde in which material absorbing at 235 nm had been removed did not give such consistent results; sometimes they were equally good and sometimes they were much more inhibitory. It is generally considered that the inhibitory effects of glutaraldehyde are related principally to material absorbing at 235 nm (Anderson, 1967). In the present study, although the most inhibitory glutaraldehyde was impure commercial glutaraldehyde which always contained much material absorbing at 235 nm, nevertheless removal of such material by charcoal washing did not always eliminate the excessive inhibitory effects. In the *Introduction* it was mentioned that some commercial glutaraldehyde had been satisfactory in the previous study

(Garrett & Harrison, 1970), using a formaldehyde-glutaraldehyde mixture, whereas new bottles of the same quality glutaraldehyde were inhibitory. On testing the old bottle it was found to contain much material absorbing at 235 nm and little or no material absorbing at 280 nm, indicating that very little actual glutaraldehyde had been present, which may have accounted for the lack of inhibition with it. From all this it would seem that on distilling the glutaraldehyde other inhibitory substances as well as material absorbing at 235 nm are removed and this suggests the possibility that some of the inhibition with non-distilled glutaraldehyde is due to inorganic contamination. An indication that this might be so was given by Fahimi & Drochmans (1968) who associated some of the inhibition of acid phosphatase with the presence of inorganic phosphates in high concentrations in commercial glutaraldehyde. In a personal communication, Roylance has indicated that in addition to phosphate, variable amounts of iron, copper and lead are sometimes present in commercially available glutaraldehyde in this country. It is, therefore, possible that inorganic contaminants in glutaraldehyde may contribute towards the inhibitory effects on enzymes.

The previous investigation (Garrett & Harrison, 1970) using immersion fixation, showed very little activity for alkaline phosphatase on the stromal surface of myoepithelial cells apart from that in sub-surface vesicles and Bogart (1968) found a similar localization in relationship to myoepithelial cells in the submandibular gland of the rat. It was wondered whether this was a real feature or the consequence of diffusion from this site plus possibly increased inhibition during the various preparative procedures. The somewhat stronger staining of the stromal surface of myoepithelial cells seen after good perfusion fixation supports these possibilities. Perfusion fixation gives more rapid uniform fixation than by immersion and thereby appears to effect a more satisfactory *in situ* immobilization of the enzyme in this situation. The constant finding of variation in the amount of reaction product around adjacent myoepithelial cells, even with the best fixation procedures, seems more likely to reflect a true concentration difference. The significance of this is not understood. It may relate to differences in the metabolic state of the cells. Bässler & Brethfeld (1968) have found that the functional state of the mammary gland of the rat affects the intensity of staining for alkaline phosphatase by the myoepithelial cells. It is not known whether the different patterns of myoepithelial cell staining frequently observed in the different glands of the same animal, with strongest reactivity in the sublingual gland and weakest in the parotid gland, reflect a true variation in concentration of the enzyme or a differential susceptibility to inhibition. The glands often showed variations in the penetration of fixatives during immersion fixation; it tended to be most uniform in the sublingual gland and least uniform in the parotid gland. Thus it seems that subtle differences in the parenchymal elements of the different glands and their milieu influence the effects of the fixation processes.

The improved preservation of enzyme reactive sites after the 'double-perfusion', using a low concentration glutaraldehyde-formaldehyde mixture for a short while and then flushing out by formaldehyde, enabled reactivity to be detected around basal duct cells and between some acinar cells for the first time at the electron microscopical level. It is considered that the glutaraldehyde enables better *in situ* immobilization of the enzyme and, if it is not allowed to continue causing too much inhibition, the paradoxical situation arises that more enzyme sites may then be detected than after the less inhibitory formaldehyde alone. This serves to emphasize that, although it may be relatively easy to say where an enzyme is present by using cytochemical methods, it is rarely possible to say where it is absent because of the possibility of diffusion away from its original site and the problem of the inhibitory effects of fixation and other parts of the procedure. In consequence, since staining has now been found between some acinar cells after the most satisfactory preservation procedure, it seems likely that the enzyme occurs more frequently in this situation *in vivo* than has been detected microscopically.

The use of DMSO is becoming popular in enzyme histochemistry for two reasons. It protects the tissue against freezing damage and it increases the permeability of the cells, thus enabling better penetration of the substrate mixtures. It has been found to be beneficial by Helminen & Ericsson (1970) for the demonstration of lysosomal acid phosphatase; leakage of enzyme from lysosomes or other enzyme-containing structures was not observed. However, in the present study, localization of the alkaline phosphatase final reaction product was more diffuse when DMSO had been used. It is wondered whether this spreading of final reaction product is connected with changes induced in the plasma-membrane of the cell by the DMSO in bringing about the increase in its permeability, and thereby some of the enzyme, which appears to be located on the outer aspect of this membrane, becomes detached. Enzymes within the cell, and especially inside lysosomes, would be more protected against such movement. In our hands, ice-crystal artifact was never a problem with tissue previously washed in a 7.5% sucrose mixture if special care were taken to freeze it very rapidly and to ensure that it was never allowed to warm above about −30°C until cut as a section.

In the past, attempts have often been made to make fixative mixtures isosmotic. However, glutaraldehyde-containing fixatives are often hypertonic and the mixture advocated by Karnovsky (1965) was found useful despite its high tonicity. Cartensen *et al.* (1971) have found that fixation rapidly increases the permeability of red cells, which are not then subject to simple osmotic stress. The tonicity effect of the fixative itself is, therefore, likely to be only short-lasting in sites that are rapidly fixed. Bohman & Maunsbach (1970) found that addition of macromolecular material, such as dextrans, was beneficial for tissue morphology using glutaraldehyde-containing fixatives by perfusion and this was attributed to its colloid osmotic effect. In the present study there was excellent tissue morphology after inclusion of dextran in the perfusion mixture and the best localization of enzyme reaction product was also achieved. Dextran was, therefore, also tested as an addition for immersion fixation and it appeared to have no adverse effects on tissue morphology and was thought to enhance the localization of enzyme reaction product. In a similar manner Ericsson *et al.* (1972) have found that fixation with glutaraldehyde mixtures made more hypertonic by the addition of sucrose improved the localization of the final reaction product for acid phosphatase. We do not necessarily advocate inclusion of dextran for all perfusion fixations for it has been found that, in Wistar strain rats, perfusion per-cardia of one-fifth strength Karnovsky's (1965) fixative containing dextran lead to gross separation of acinar cells in the parotid glands (Garrett & Parsons, unpublished observations). This separation of cells was not conspicuous when dextran was included in the immersion fixation of parotid tissues from similar rats. It is wondered whether the untoward perfusion effect is attributable to a massive histamine release, for such an effect has been observed to occur very rapidly in albino rats perfused with dextran solutions (Halpern, 1956).

Nothing is known about the possible function(s) of alkaline phosphatase(s) in salivary glands. It has been confirmed that the salivary myoepithelial cells in the cat are capable of contraction by Emmelin *et al.* (1968) but the role of the basal duct cells is unknown. Hamperl (1970) included basal duct cells under the heading 'myothelia' but the description of basal duct cells in the submandibular gland of the cat by Shackleford & Wilborn (1970) indicates that they are not very similar to myoepithelial cells. Not all myoepithelial cells in all species possess cytochemically detectable alkaline phosphatase (Garrett & Harrison, 1970) and thus it does not appear to be essential for the function of all myoepithelial cells. Alkaline phosphatase has now been found between some salivary acinar cells in the cat and around the lumen of some acini-sites where it is abundantly present in the sublingual gland of the dog (Garrett & Harrison, 1970). A possible role for the enzyme may be in transport mechanisms of some organic compounds in or out of cells involving a hydrolysis of their phosphate esters. This may be to provide essential metabolites for some cells or to assist in transport between adjacent cells and possibly in secretory processes. More work is required to get even a glimpse of any true function of alkaline phosphatase in salivary glands.

References

ANDERSON, P. J. (1967). Purification and quantitation of glutaraldehyde and its effect on several enzyme activities in skeletal muscle. *J. Histochem. Cytochem.* **15**, 652–61.

BÄSSLER, R. & BRETHFELD, V. (1968). Enzymhistochemische Studien an der Milchdrüse. *Histochemie* **15**, 270–86.

BOGART, B. I. (1968). The fine structural localization of alkaline and acid phosphatase activity in the rat submandibular gland. *J. Histochem. Cytochem.* **16**, 572–81.

BOHMAN, S-O. & MAUNSBACH, A. B. (1970). The effects on tissue fine structure of variations in colloid osmotic pressure of glutaraldehyde fixatives. *J. ultrastruct. Res.* **30**, 195–208.

CARSTENSEN, E. L., ALDRIDGE, W. G., CHILD, S. Z., SULLIVAN, P. & BROWN, H. H. (1971). Stability of cells fixed with glutaraldehyde and acrolein. *J. Cell Biol.* **50**, 529–32.

EMMELIN, N., GARRETT, J. R. & OHLIN, P. (1968). Neural control of salivary myoepithelial cells. *J. Physiol., Lond.* **196**, 381–96.

ERICSSON, J. L. E., BRUNK, U. T. & FORSBY, N. H. (1972). On the demonstration of acid phosphatase in tissue cultured cells. Studies on the significance of fixation, tonicity and permeability factors. *Proc. R. micr. Soc.* **7**, 113–14.

FAHIMI, H. D. & DROCHMANS, P. (1968). Purification of glutaraldehyde. Its significance for preservation of acid phosphatase activity. *J. Histochem. Cytochem.* **16**, 199–204.

FRIGERIO, N. A. & SHAW, M. J. (1969). A simple method for determination of glutaraldehyde. *J. Histochem. Cytochem.* **17**, 176–81.

GARRETT, J. R. & HARRISON, J. D. (1970). Alkaline-phosphatase and adenosinetriphosphatase histochemical reactions in the salivary glands of cat, dog and man, with particular reference to the myoepithelial cells. *Histochemie* **24**, 214–29.

HALPERN, B. N. (1956). Histamine release by long chain molecules. In *Ciba Foundation Symposium on Histamine* (G. E. W. Wolstenholme & C. O. O'Connor, eds), pp. 92–123. London: Churchill.

HAMPERL, H. (1970). The Myothelia (myoepithelial cells). *Current Topics in Path.* **53**, 161–220.

HELMINEN, H. J. & ERICSSON, J. L. E. (1970). On the mechanism of lysosomal enzyme secretion. Electron microscopical and histochemical studies on the epithelial cells of the rat's ventral prostate lobe. *J. ultrastruct. Res.* **33**, 528–49.

HOPWOOD, D. (1970). The relations between formaldehyde, glutaraldehyde and osmium tetroxide, and their fixation effects on bovine serum albumin and tissue blocks. *Histochemie* **24**, 50–64.

HUGON, J. & BORGERS, M. (1966*a*). A direct lead method for the electron microscopic visualization of alkaline phosphatase activity. *J. Histochem. Cytochem.* **14**, 429–31.

HUGON, J. & BORGERS, M. (1966*b*). Ultrastructural localization of alkaline phosphatase activity in the absorbing cells of the duodenum of mouse. *J. Histochem. Cytochem.* **14**, 629–40.

KARNOVSKY, M. J. (1965). A formaldehyde-glutaraldehyde fixative of high osmolality for use in electron microscopy. *J. Cell Biol.* **27**, 137–38A.

SHACKLEFORD, J. M. & WILBORN, W. H. (1970). Ultrastructural aspects of cat submandibular glands. *J. Morph.* **131**, 253–76.

STUTTE, H. J. (1967). Hexazotiertes Triamino-tritolyl-methanchlorid (Neufuchsin) als Kupplungs-salz in der Fermenthistochemie. *Histochemie* **8**, 327–31.

Fixation and tissue preservation for antibody studies

H. R. P. MILLER

Department of Veterinary Pathology,
University of Glasgow, Scotland

Contents

Synopsis. The methods of fixation and preparation of lymphoid tissues for the immuno-enzyme technique are reviewed. For this technique an enzyme is used first as an antigen and then as a marker to demonstrate its specific antibody. A variety of commonly employed fixatives satisfactorily conserve tissues for the light microscopic detection of antibody but, for electron microscopy, glutaraldehyde or formaldehyde or both are the fixatives of choice. The main technical problem for electron microscopy is to reduce the size of the tissue fragments sufficiently so that the enzymes and their substrates permeate through the fixed tissues. The merits and short-comings of the different preparative techniques are examined and it is shown that the most reproducible results are obtained with 40 μm frozen sections. Some of the problems of non-specific staining arising from fixation procedures, as well as endogenous enzyme activity, are discussed. The evidence for and against antibody inactivation by fixation and enzyme inactivation by interaction with its specific antibody is reviewed.

Introduction

Antibody-forming cells can be detected by both light and electron microscopy using the immuno-enzyme technique developed by Avrameas and his co-workers (Avrameas & Lespinats, 1967; Avrameas, 1970; Leduc *et al.*, 1968; Scott *et al.*, 1968). With this method an enzyme is first used as an antigen and, subsequently, as a marker to demonstrate its specific intracellular antibodies.

The technique has several advantages over other systems, notably, that a simple optical microscope may be used, that the preparations are permanent and that intracellular distribution of antibody can be studied ultrastructurally. It is also possible to demonstrate different anti-enzyme antibodies in the same preparation (Avrameas *et al.*, 1971) or the relationship of antibody-forming cells to lysosome-bearing macrophages (Straus, 1970*a*).

Briefly, the technique requires that the tissues are first fixed and are then reduced in size or in thickness before they are incubated in enzyme so that the latter can diffuse into all the cells. The specimen is then carefully washed to eliminate unattached enzyme and is incubated with an appropriate substrate to reveal the sites of enzyme (antibody) activity.

Several problems arise when considering the types of fixative and the methods of tissue preparation that may be used to demonstrate anti-enzyme antibodies. In order to retain morphological detail, the tissues must be fixed before they are treated with enzymes and substrates. However, fixation should not significantly reduce the activity of the antibody, nor should it predispose to non-specific tissue staining. For electron microscopy, the methods of preparing the fixed tissues before their incubation in enzyme-containing media influence the diffusion of both the enzymes and of the substrates into the cells.

In this paper, the methods of fixation and of processing the tissues employed to demonstrate anti-enzyme antibodies are reviewed. Some of the problems concerning the quality of fixation, the diffusion of enzyme and substrate into the tissues, and of non-specific staining are also discussed.

Enzymes as antigens and markers

In theory, any enzyme that is not toxic, that is histochemically demonstrable, and which provokes an immune response can be employed. It should not, however, change the structure of the cell or of the tissues. In practice, enzymes of relatively low mol. wt. are preferred, especially for electron microscopy, since they diffuse more readily into fixed tissues. It is also important to use the purest enzyme preparation available and one that has a high specific activity.

Avrameas & Lespinats (1967) showed that horseradish peroxidase, acid phosphatase from potato, and *Aspergillus niger* glucose oxidase could be used to demonstrate antibody-forming cells by light microscopy. Subsequently, horseradish peroxidase (Leduc *et al.*, 1968; Avrameas & Leduc, 1970), *E. coli* alkaline phosphatase (Scott *et al.*, 1968), glucose oxidase, and *Dactylium dendroides* galactose oxidase (Kuhlmann & Avrameas, 1971) have been employed to localize antibody ultrastructurally.

A variety of substrates are available for the cytochemical detection of enzymes. For example, Straus (1970*b*) used 3-amino-9-ethyl-carbazole and hydrogen peroxide to mark the site of peroxidase antigen with a red reaction product and, after treating the tissues with enzyme, he used benzidine and hydrogen peroxide to give a blue final reaction product at the sites of antibody synthesis. 3,3′-diaminobenzidine (Graham & Karnovsky, 1966) is used as the substrate (with hydrogen peroxide) for electron microscopy and is also highly satisfactory for light microscopy (Figs. 1, 2).

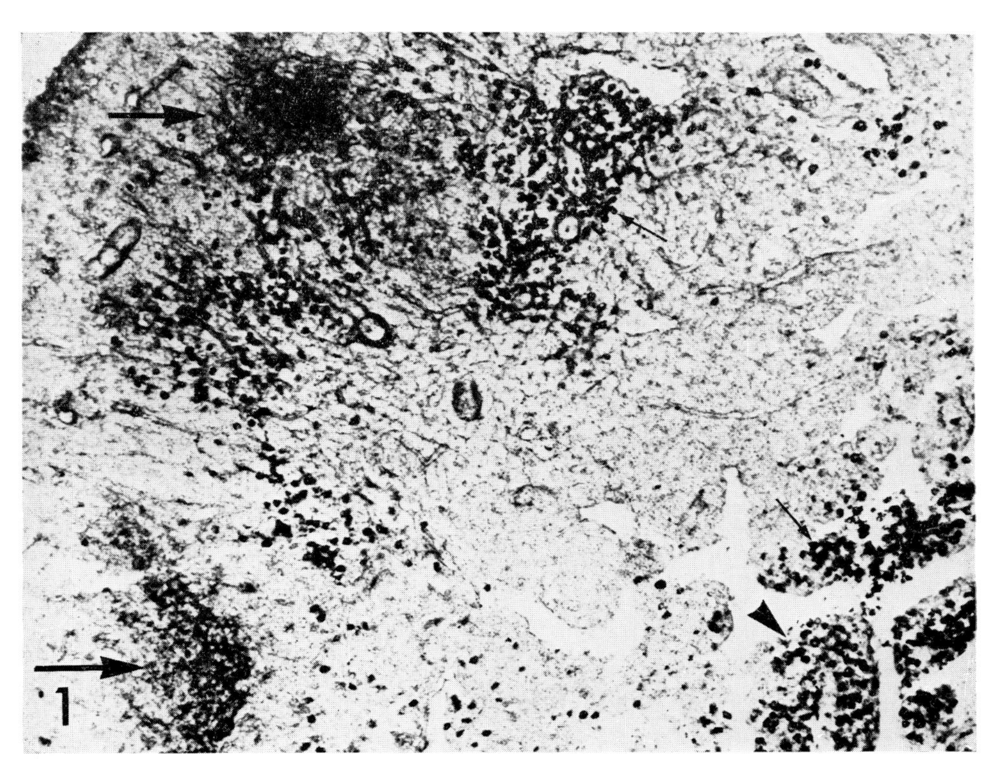

Figure 1. Low power photomicrograph of a 40 μm frozen section from mouse popliteal lymph node. The mouse was immunized in the hind foot-pad with horseradish peroxidase in Freund's complete adjuvant 14 days previously. Two germinal centres (large arrows) and the medullary cords (arrowhead) are visible. Numerous antibody-forming cells (small arrows) are present in the medullary cords and in the region of the germinal centre. Fixation, 4% formaldehyde for 24 hr. × 40

Similarly, there are several substrates available for the light microscopic detection of alkaline and acid phosphatases (Avrameas, 1970). The substrate β-glycerophosphate with lead nitrate at pH 8 (Hugon & Borgers, 1966) is used for the ultrastructural localization of alkaline phosphatase (Scott *et al.*, 1968; Miller & Avrameas, 1971) but is unsatisfactory for light microscopy. Procedures for the cytochemical and ultrastructural demonstration of glucose and galactose oxidases have also been described recently (Avrameas, 1970; Kuhlmann & Avrameas, 1971). Additionally, by making use of the different colours produced by the substrates for different enzymes, it is practicable to detect different antibodies synthesized in the same tissues (Avrameas *et al.*, 1971).

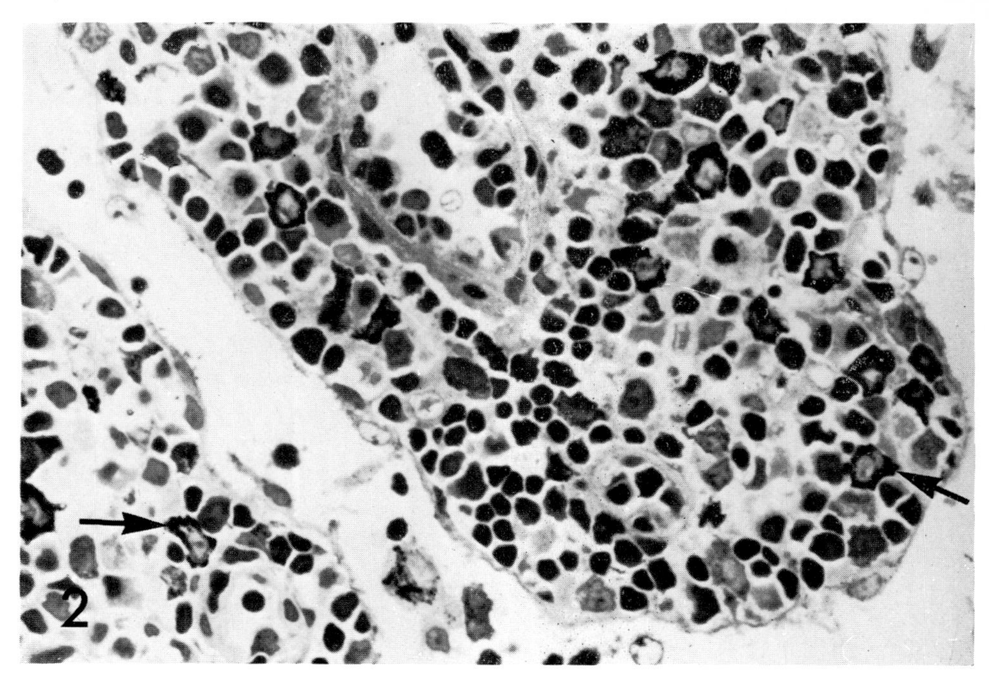

Figure 2. 1 μm section from a 40 μm frozen section of a node 3 days after secondary stimulation with horseradish peroxidase. Antibody-forming cells (arrows) are abundant in the medullary cord. The section was counterstained with Azure II-Methylene Blue to demonstrate the cytology of the cells. Fixation as in Fig. 1. × 500

Fixation of tissues

Several problems arise when considering which fixative is best employed because, although it is desirable to have satisfactory conservation of the tissues, fixation should not reduce the activity of the antibody nor should it impede the diffusion of the enzyme and substrate into the tissues and cells.

For light microscopy, imprints, cell suspensions, or frozen tissue sections can be fixed with acetone, absolute methanol, 95% alcohol, and alcohol-ether (60:40 v/v) (Avrameas, 1970). All of the above-mentioned enzymes are demonstrable when alcohol-ether has been used as a fixative (Avrameas & Lespinats, 1967; Kuhlmann & Avrameas, 1971). However, the conservation of frozen tissue sections may be improved where the specimens are fixed before being sectioned. Straus (1968) studied several different fixation methods for the demonstration of anti-horseradish peroxidase antibody in the spleen and lymph nodes of rabbits. He concluded that the best results were obtained with freshly prepared 4% formaldehyde and a fixation time of 24 hr. Similarly, Kuhlmann & Miller (1971) found that antibody forming cells were satisfactorily conserved and readily demonstrable by light microscopy with this fixation procedure.

For ultrastructural studies, the parameters are more critical because even small amounts of antibody diffusion, or of non-specific staining, are at once evident. Glutaraldehyde and a formaldehyde-glutaraldehyde mixture have both been employed for the ultrastructural localization of anti-horseradish peroxidase (Leduc *et al.*, 1968; Sordat *et al.*, 1970; Kuhlmann & Miller, 1970). Glutaraldehyde is also satisfactory for the demonstration of anti-alkaline phosphatase antibody (Scott *et al.*, 1968). Fixation times of up to 1.5 hr using a concentration of 2.5% glutaraldehyde did not appear to significantly decrease the binding of enzymes in the tissues (Kuhlmann & Miller, 1971). However, unless perfusion fixation is carried out, only relatively small blocks of tissues can be fixed with glutaraldehyde.

Leduc *et al.* (1969) recommended the use of low concentrations of formaldehyde and employed a 1% solution for 30 min to 1 hr. They found that the addition of 0.25 M sucrose prevented diffusion of the antibody. The same fixation procedure was subsequently used by Avrameas & Leduc (1970) to demonstrate anti-horseradish peroxidase antibody in two different cell types. Kuhlmann & Miller (1971) tried varying concentrations of formaldehyde over different time intervals and found that 24 hr fixation in 4% formaldehyde improved tissue preservation and did not appear to inhibit the subsequent reaction of the enzymes with their specific antibody. For adequate fixation with formaldehyde, immersion for a minimum of 4 hr was necessary and antibody activity was maintained even after very long periods of fixation (Kuhlmann & Miller, 1971). There are three main advantages in using formaldehyde as compared with glutaraldehyde.

(1) Larger blocks can be fixed by immersion so that the entire structure of a small lymph node can be preserved.
(2) There is less background staining with horseradish peroxidase using formaldehyde and 40 μm sections can be examined by light microscopy.
(3) Ultrastructurally, there is less contrast in the cell membranes, and thus the early stages of antibody synthesis can be more readily detected.

However, with glutaraldehyde the plasmalemmata and Golgi complexes are better conserved and this is particularly noticeable in tissues which have been treated with diaminobenzidine.

Preparation of tissues for incubation with antigen

For light microscopy, the commonly employed method of preparing tissues for the immuno-enzyme technique is to cut frozen sections (Straus, 1968, 1970*a*,*b*; Miller, 1970; Kuhlmann & Miller, 1971). Sordat *et al.* (1970) used freeze-substitution and polyester-wax embedding with equal success. The possibility of using formalin fixation and cold paraffin embedding techniques has not yet been explored. The best results are obtained with 1 μm plastic sections cut from Epon-embedded 40 μm frozen sections treated to reveal anti-horseradish peroxidase antibody (Fig. 2), although this technique has not yet been applied to demonstrate anti-alkaline phosphatase antibody.

For electron microscopy there are two different approaches. The principal difficulty is in reducing the size of the tissues sufficiently that both the enzymes and their substrates can reach all the cells. The first approach is to cut the fixed tissue blocks into small fragments with a razor-blade (Leduc *et al.*, 1968) which causes less cell damage than a mechanical tissue chopper (Kuhlmann & Miller, 1971). It should, however, be pointed out that the main tissues studied have been lymph node and spleen which, structurally, may be more susceptible to damage than other tissues.

The fragments are incubated with enzyme (1 mg/ml) for 1 to 24 hr (Leduc *et al.*, 1968; Kuhlmann & Miller, 1971) and are washed carefully to remove excess unfixed enzyme. They may then be briefly refixed in either glutaraldehyde (Leduc *et al.*, 1968) or formaldehyde (Kuhlmann & Miller, 1971) to minimize the diffusion of enzyme. This latter step is not abolutely necessary, although Kraehenbuhl *et al.* (1971) found that it reduced the damage to the cell membranes following treatment with diaminobenzidine. After this secondary fixation, the fragments are washed, treated with the appropriate substrate, and post-fixed in osmium tetroxide. These small tissue fragments, although they show the best ultrastructural preservation of the cells, have several disadvantages:

(1) There is poor penetration of both enzymes and substrates (see later);
(2) it is not possible to demonstrate anti-alkaline phosphatase antibody (Kuhlmann & Miller, 1971); and
(3) the relation of each individual fragment to the structure of the node as a whole is difficult to assess.

The second approach is to cut frozen sections of already fixed lymph node and both anti-alkaline phosphatase and anti-horseradish peroxidase antibodies have been demonstrated in ultra-thin frozen sections (Scott & Avrameas, 1968; Kuhlmann & Miller, 1971). However, this technique is, at present, of limited application. A simple approach is to cut 40 μm frozen sections in an ordinary cryostat (Miller, 1970; Kuhlmann & Miller, 1971). To prevent ice-crystal formation, the tissue blocks are soaked in 10% dimethyl sulphoxide for 1 hr before freezing and cutting (Miller, 1970; Kuhlmann & Miller, 1971). The tissue blocks are then frozen with carbon dioxide or in liquid nitrogen and cut at -20°C. The sections are dropped immediately into buffer at room temperature (Miller, 1970; Kuhlmann & Miller, 1971) and are subsequently treated in the same way as the small tissue fragments to reveal antibody activity. Once the sections have been impregnated with Epon, they are embedded flat so that they can be viewed by light microscopy and it is then possible to select specific areas in the node. Frozen sections have an added advantage that all four enzymes are demonstrable (Scott *et al.*, 1968; Kuhlmann & Miller, 1971; Kuhlmann & Avrameas, 1971).

Interpretation of artefacts

(a) *Non-specific staining*

Some enzymes have an endogenous activity in the tissues; for example, peroxidase is present in red blood cells and in leukocytes, and acid phosphatase is found in the lysosomes of macrophages. Endogenous enzyme activity, as well as antigen, can be distinguished from antibody by omitting incubation in enzyme and treating the tissues with substrate. Non-specific attachment of the enzyme is detected by incubating tissues, immunized with a heterologous antigen, in enzyme and then staining with the appropriate substrate. In addition, the specificity of the substrate can be tested by omitting essential ingredients such as hydrogen peroxide from the substrate for horseradish peroxidase or β-glycerophosphate from Hugon and Borgers' medium for alkaline phosphatase. In the latter instance there is, at pH 8, a non-specific fine lead deposit on the cell membranes.

After glutaraldehyde fixation, there is considerable background staining with horse-

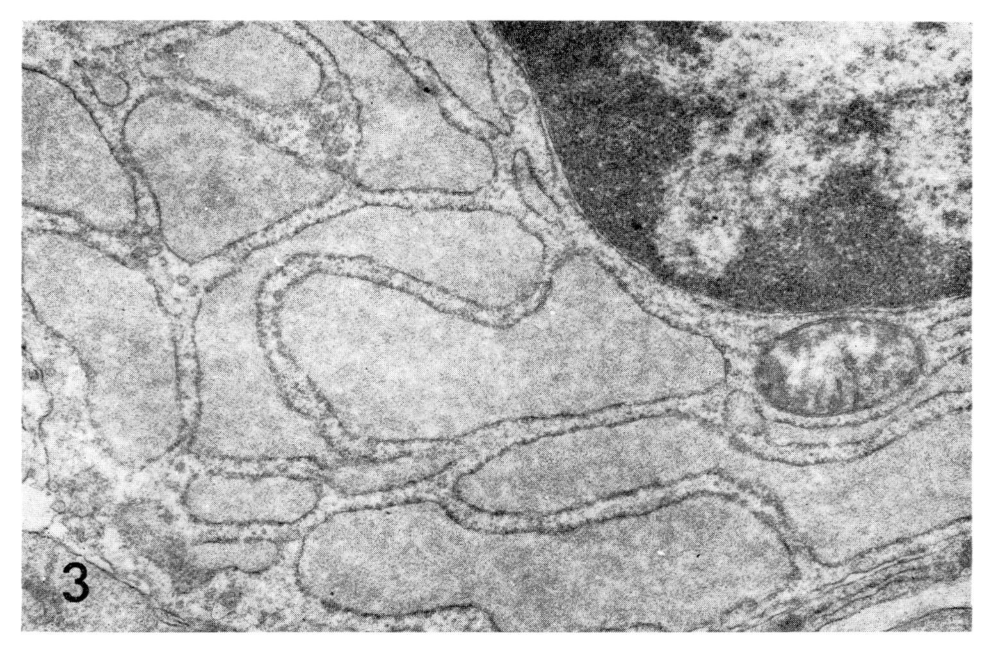

Figure 3. Electron micrograph of a plasma cell from a control mouse immunized with alkaline phosphatase but treated with diaminobenzidine to reveal peroxidase activity. Note the contrast of the membranes of the rough endoplasmic reticulum and of the nuclear chromatin even though the thin section was not counterstained. Fixation, 2.5% glutaraldehyde for 1.5 hr. Preparation, 40 μm frozen section. × 28 000

radish peroxidase and it is difficult to distinguish both the antibody-forming cells and the structural features of the node in 40 μm frozen sections. Glutaraldehyde fixation and subsequent treatment of the tissues with diaminobenzidine also increases the contrast of membranes at the ultrastructural level (Fig. 3). However, after formaldehyde fixation, background staining in 40 μm sections is minimal (Fig. 1) and, ultrastructurally, the membrane contrast is very low (Fig. 4).

With horseradish peroxidase and other enzymes where diaminobenzidine is the substrate, the thin sections should be examined initially without counter-staining because, in the early stages of antibody synthesis, only the membranes of the rough endoplasmic reticulum may show antibody activity. Membrane contrast is enhanced after counter-staining for 30 sec with lead citrate and this tends to obscure the weak antibody staining of the rough endoplasmic reticulum. Counter-staining with uranyl acetate and lead citrate is useful where cytological details are to be examined (Fig. 5).

(b) *Diffusion of enzyme and substrate into tissues*

The penetration of enzyme and substrate into tissue fragments is very limited (Leduc *et al.*, 1968; Sordat *et al.*, 1970), although fixation with formaldehyde and long incubation periods in enzyme can result in uniform staining of some fragments. Furthermore, whilst there is no dubiety when cells are positive for antibody, it is not certain if cells are antibody-free merely because the enzymes and their substrates have not diffused sufficiently deep into the tissue fragments.

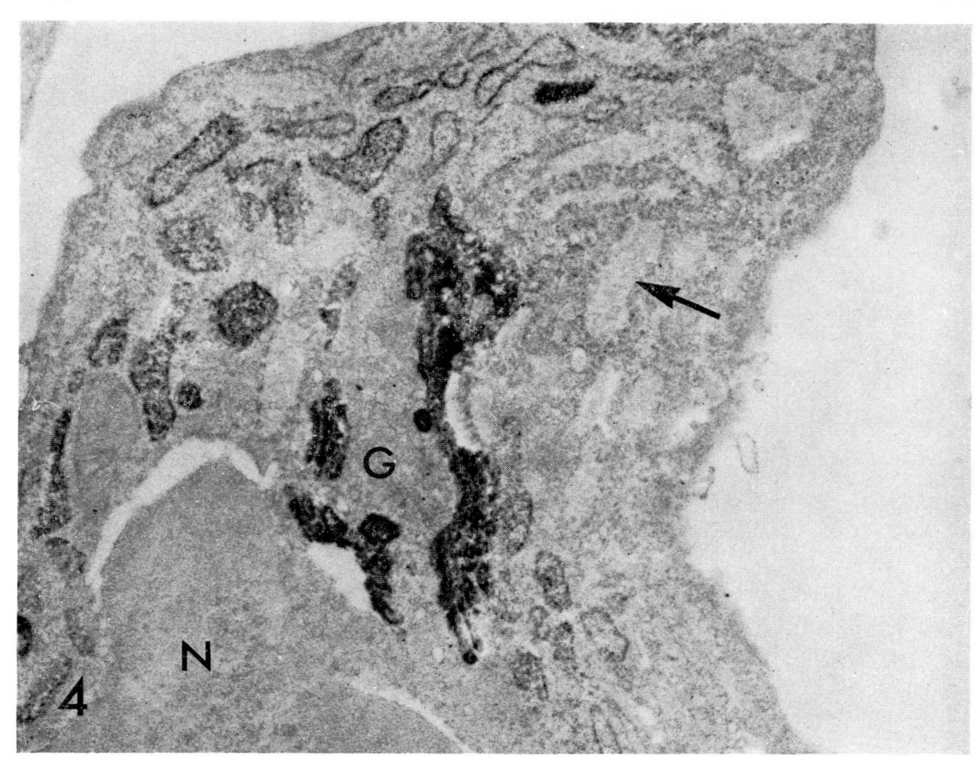

Figure 4. In this plasma cell there is accumulation of anti-horseradish peroxidase antibody in the Golgi complex (G) and in some cisternae of rough endoplasmic reticulum. Note that in this section, there is low membrane contrast in the antibody-free cisternae of the rough endoplasmic reticulum (arrow) and the nuclear chromatin (N) is not visible (compare with Fig. 3). Ten days after primary immunization with horseradish peroxidase. Fixation and preparation as in Fig. 1. The section was not counterstained. × 30 000

The maximum distance that enzymes and their substrates must penetrate into frozen sections is 20 μm and differential cell counts do not show any differences in the staining intensities of cells at the centres or towards the surfaces of the sections (Fig. 6). This would suggest that, unlike tissue fragments, no diffusion gradient is present in the sections. Kuhlmann & Miller (1971) also observed that intracellular penetration of enzymes

---

Figure 5. Plasma cell synthesizing anti-horseradish peroxidase antibody 6 months after primary stimulation. Reaction product is concentrated in the Golgi complex (G) and in many cisternae of the rough endoplasmic reticulum. The plasmalemmata are damaged at several points (arrow). Fixation, 2.5% glutaraldehyde for 1.5 hr. Preparation, 40 μm frozen section. Uranyl acetate and lead citrate counterstaining. × 10 000

Figure 6. Light micrograph of a 1 μm cross section from a 40 μm frozen section. The cells at the centre of the section stain as intensely as those at the surfaces. Specific antibody reaction is present in the Golgi regions of most of the antibody-forming cells (arrows). From the medullary region of a mouse popliteal node 20 days after stimulation with horseradish peroxidase. Fixation and preparation as in Fig. 1. The section was not counterstained. × 1000

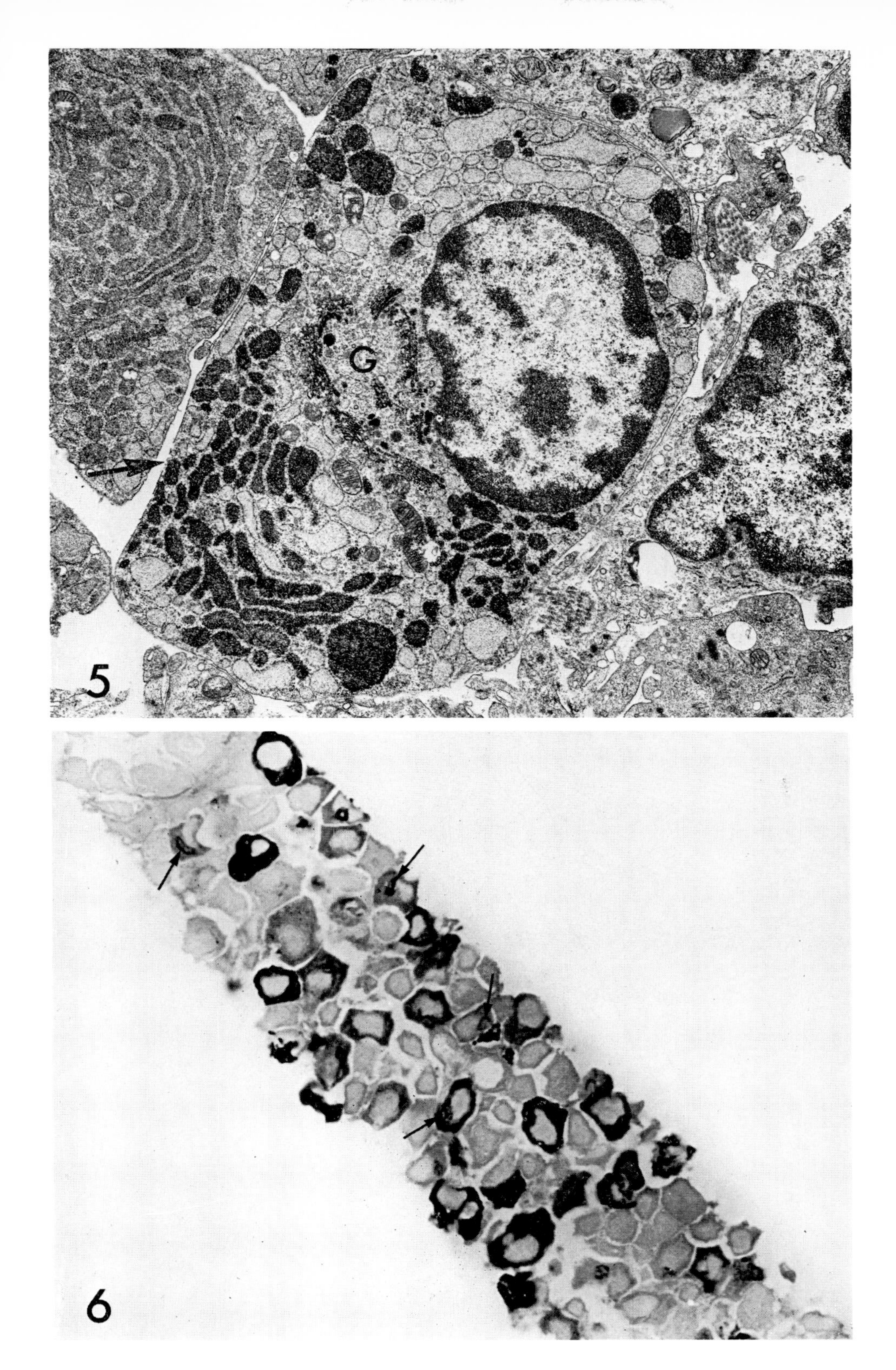

5
G
6

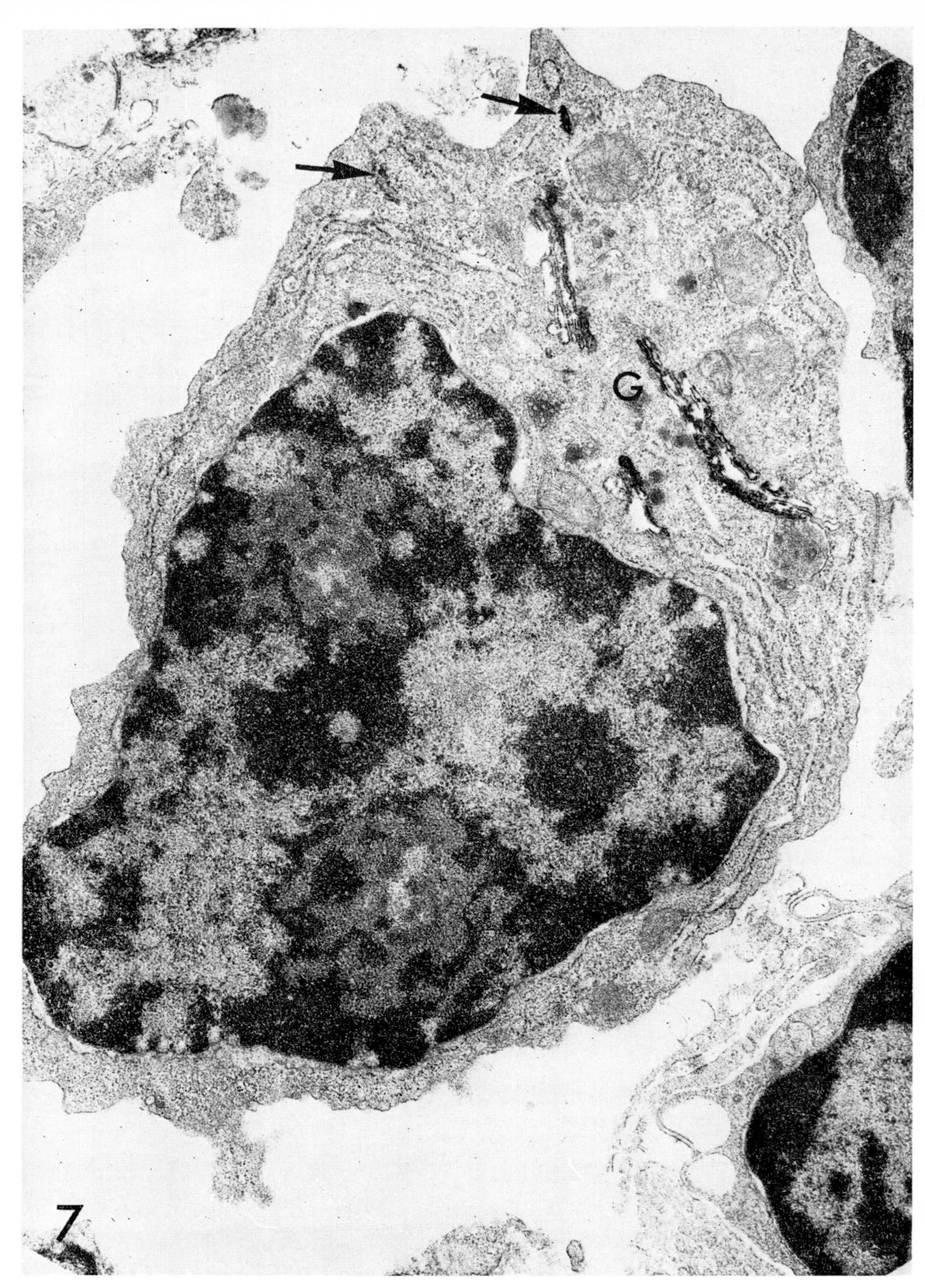

Figure 7. Electron micrograph of a plasma cell 8 days after primary stimulation with horseradish peroxidase. The localization of antibody in this cell is typical of the distribution seen in immunocytes early in the first response (Miller *et al.*, 1972). Antibody is concentrated in the Golgi region (G) and small amounts are present in a few cisternae of the rough endoplasmic reticulum (arrows). Fixation and preparation as in Fig. 1. Section counterstained with uranyl acetate and lead citrate. × 20 000

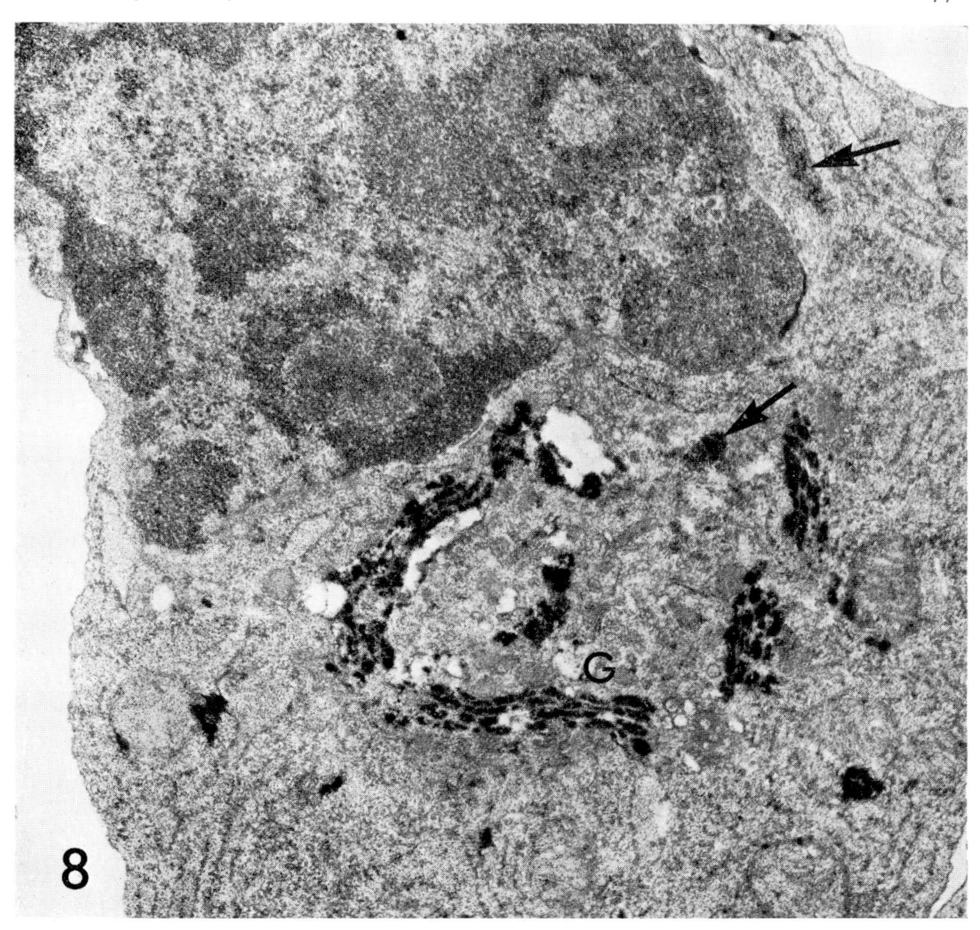

Figure 8. Anti-alkaline phophatase antibody is concentrated in the Golgi complex (G) of this plasma cell early in the primary response. Some antibody is also present in a few cisternae of the rough endoplasmic reticulum (arrows). The distribution of anti-alkaline phosphatase antibody is similar to that of anti-horseradish peroxidase and, in the primary response, there is usually an early concentration of anti-alkaline phosphatase antibody in the Golgi complex before the cisternae of the rough encoplasmic reticulum are filled (Miller, unpublished observations). Fixation and preparation as in Fig. 1. Uranyl acetate and lead citrate counterstaining. × 32 000

was apparently superior in frozen sections because a greater proportion of cells showed intense antibody activity in their Golgi complexes. This was tested by comparing frozen sections and tissue fragments taken from the same popliteal nodes of five mice, 20 days after immunization with horseradish peroxidase. 62% (S.E.M. ±5) of the antibody-forming cells from frozen sections had antibody in their Golgi regions (Fig. 6), whereas in tissue fragments 15% (S.E.M. ±3) had stained Golgi regions and all were antibody-forming cells at the periphery of the blocks. Ultrastructurally, the difference is clearly illustrated and shows the early accumulation of antibody in the Golgi complexes of antibody-forming cells in frozen sections (Figs. 4, 7 & 8) whereas the majority of antibody-forming cells in tissue fragments rarely have antibody in this site (Fig. 9).

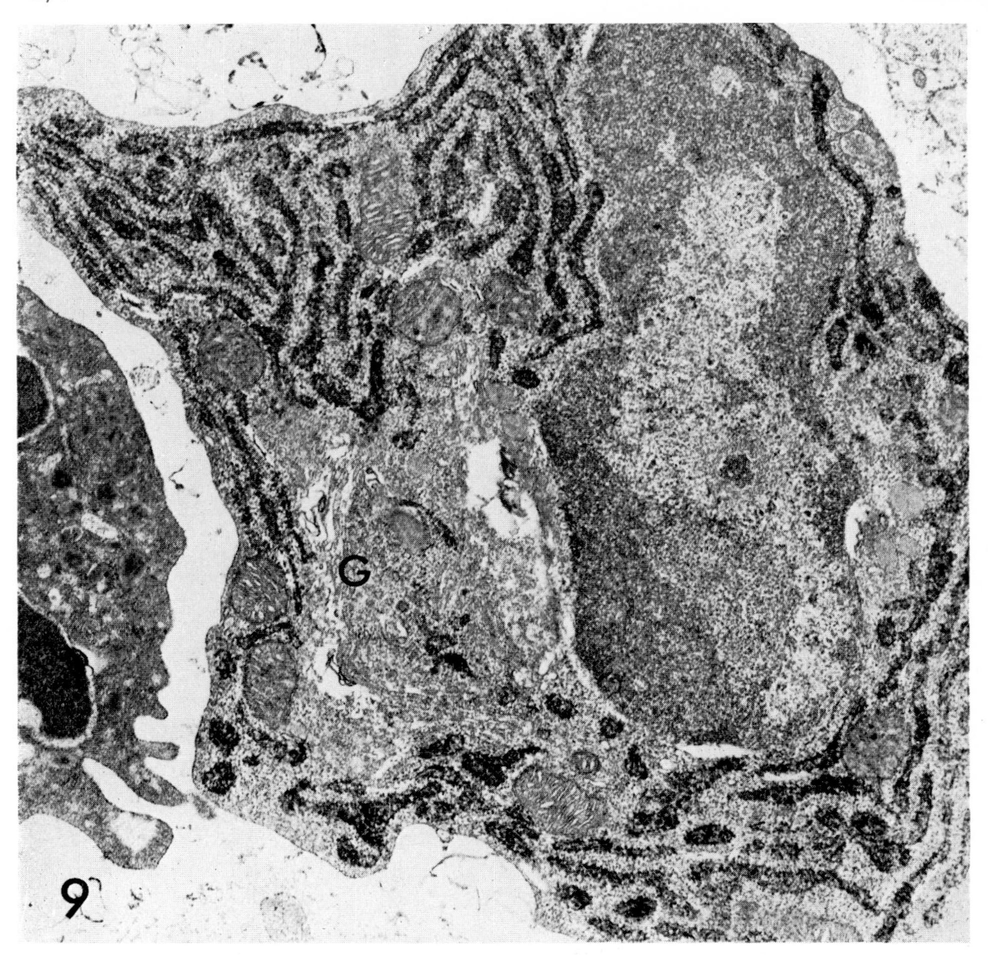

Figure 9. Plasma cell synthesizing antibody 20 days after primary stimulation. All of the cisternae of the rough endoplasmic reticulum contain anti-horseradish peroxidase antibody but the Golgi complex (G) is empty and has dilated cisternae. Fixation as in Fig. 1; preparation, small tissue fragment. Counterstained with uranyl acetate and lead citrate. × 17 000

Similar detailed studies have not yet been carried out with the other enzymes, although the pattern of antibody localization and of Golgi staining with alkaline phosphatase is very similar to that observed with horseradish peroxidase (Fig. 8) (Miller, unpublished observations).

(c) *Cell preservation and antibody diffusion*

Glutaraldehyde or formaldehyde-glutaraldehyde fixation of tissues and the subsequent preparation of small fragments provides the best cell preservation both for light and electron microscopy (Leduc *et al.*, 1968; Kuhlmann & Miller, 1971). Cell membranes and organelles remain intact and intracellular diffusion of antibody is seen in only a few cells which, presumably, have reached the end of their life span (Leduc *et al.*, 1968).

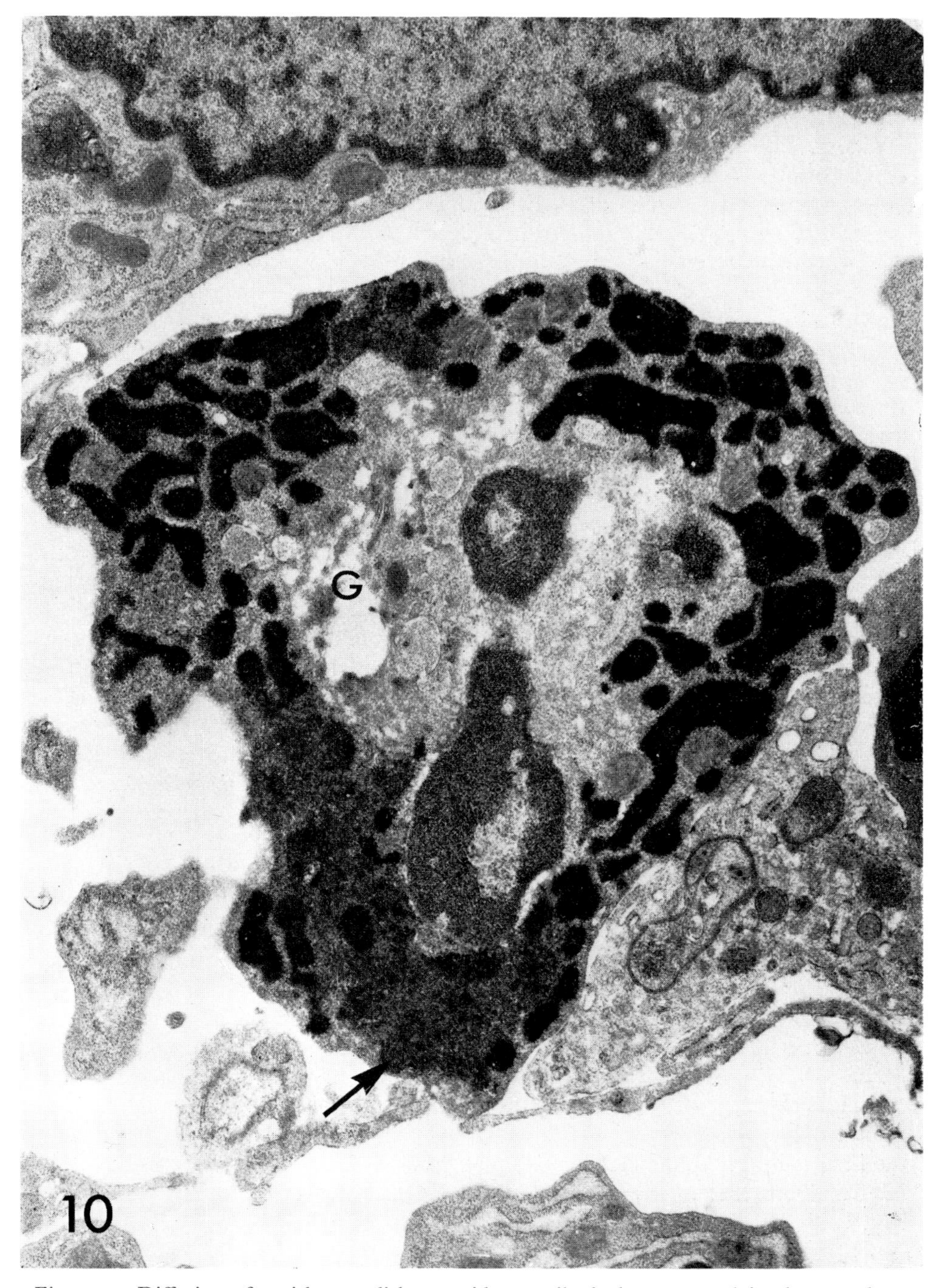

Figure 10. Diffusion of anti-horseradish peroxidase antibody has occurred in the cytoplasm (arrow) of this plasma cell. The Golgi complex (G) does not contain antibody and some of its cisternae are dilated. Fixation and preparation as in Fig. 1. Section counterstained with uranyl acetate and lead citrate. × 17 000

Where 40 μm frozen sections are cut from glutaraldehyde-fixed tissues, the localization of antibody is also excellent (Fig. 5); however, small defects are often present in the plasmalemmata of some cells (Fig. 5).

After fixation in formaldehyde, the conservation of cell membranes in tissue fragments is less satisfactory, and the Golgi cisternae are often dilated (Figs. 9 & 10), especially after diaminobenzidine treatment. The membrane damage is more severe in frozen sections and, additionally, in some sections intracellular diffusion of anti-horseradish peroxidase and anti-alkaline phosphatase antibodies may occur in up to 20–30% of the antibody-forming cells (Fig. 10). Antibody diffusion is particularly noticeable in tissues where the antibody-forming cells contain relatively high concentrations of antibody.

(d) *Antibody and enzyme inactivation*

At present, there is no entirely satisfactory method of assessing the extent of antibody inactivation following fixation. Kraehenbuhl & Campiche (1969) and Kraehenbuhl *et al.* (1971) used immuno-diffusion plates and suggested that glutaraldehyde caused greater inactivation of antibody than formaldehyde. However, immobilization of the antibody by fixation will, in itself, alter the pattern of antigen diffusion from the central well without necessarily causing a reduction of antibody activity. Since glutaraldehyde is used with great success to prepare immuno-adsorbents (Avrameas & Ternynck, 1969) it is unlikely that fixation causes any gross inhibition of antibody activity. Our own observations also have failed to demonstrate any significant difference in the staining intensity of antibody when the results of fixation with glutaraldehyde and formaldehyde are compared (Kuhlmann & Miller, 1971).

Straus (1968) reported that the binding of antibody to horseradish peroxidase caused approximately 60% inhibition of enzyme activity. In contrast, Sordat *et al.* (1970) and Hess *et al.* (1971) found that there was virtually no inhibition of horseradish peroxidase activity by either rabbit or mouse antisera. They attributed the high degree of inhibition reported by Straus to the loss of enzyme following the dilution steps carried out during the assay system. More recent work suggests that there is a small degree of inhibition; for example, with mouse antisera there is approximately 10% inhibition of horseradish peroxidase activity, and this varies according to the species tested (Avrameas, personal communication, 1972). The extent of inhibition of other enzymes by their specific antisera has not yet been examined.

It seems unlikely that either antibody inactivation by fixation or enzyme inhibition by antibody significantly interfere with the sensitivity of the immuno-enzyme technique. Moreover, the sensitivity can be enhanced when the continued catalytic activity of the enzyme is made use of by prolonging the incubation time with substrate to increase the amount of reaction product. In addition, recent studies by Hay *et al.* (1972) suggest that the immuno-peroxidase technique is more sensitive than plaque assay methods and immunofluorescence.

Conclusions

Several different fixatives may be employed for cell suspensions, smears and imprints, although the best results with frozen sections are achieved when tissues are fixed with formaldehyde before being cut in a cryostat. Since these preparations are no more than one cell thick, the enzymes and substrates do not have to diffuse to any significant depth to reach the sites of antibody synthesis. Thus, there are relatively few technical problems associated with the light microscopic detection of anti-enzyme antibodies.

For electron microscopy, glutaraldehyde and formaldehyde-glutaraldehyde fixation provide the best conservation of cell structure and prevent intracellular diffusion of antibodies even in frozen sections. However, when horseradish peroxidase is used with glutaraldehyde-fixed tissues, there is intense background staining in 40 μm frozen sections and, when diaminobenzidine is employed as substrate, there is an increased membrane contrast. Cell structure is less well preserved after formaldehyde fixation, especially in frozen sections where intracellular diffusion of antibody may occur, but background staining and membrane contrast are negligible.

To achieve uniform diffusion of enzyme and substrates into the fixed tissues they are either sliced into tiny fragments with a razor blade, or are cut in a cryostat. The cell morphology is better preserved in fragments but there is variable diffusion of enzymes and substrates into the tissues and the intracellular penetration is poor. In 40 μm frozen sections, there is membrane damage in both glutaraldehyde- and formaldehyde-fixed cells and, with the latter fixative, intracellular diffusion of antibody occurs. However, the enzymes and substrates permeate readily through the sections and the intracellular penetration is good.

It is clear from these observations that, for detailed studies of antibody synthesis and secretion, several different fixatives and both techniques for tissue preparation should be employed. Fixation with formaldehyde and 40 μm frozen sections are preferable for semi-quantitative studies of the immune response, because selected areas in the nodes can be repetitively sampled. Moreover, this method is more easily controlled since alternate sections can be used for experimental and control purposes.

The direct immuno-enzyme technique, whilst of obvious interest to the immunologist, may also be a useful experimental model for the cell biologist. Because the rough endoplasmic reticulum and the Golgi complex are both involved in the synthesis and secretion of antibody, the technique provides a unique opportunity to visualize the passage of protein from the rough endoplasmic reticulum into the Golgi region and to study the relationship between these two organelles.

Acknowledgements

Much of this work was done in collaboration with Drs W. Kuhlmann, S. Avrameas and E. H. Leduc while the author was on study leave from the University of Glasgow and was supported by a grant from the Carnegie Trust for the Universities of Scotland. I am grateful to Dr W. Bernhard and Professor W. F. H. Jarrett for their generous provision of facilities, to Mr J. Morrison and Mr A. Finnie for preparing the micrographs, and to Mrs E. Paterson for typing the manuscript.

References

AVRAMEAS, S. (1970). Immuno-enzyme techniques: enzymes as markers for the localisation of antigens and antibodies. *Int. Rev. Cytol.* **27**, 349–85.

AVRAMEAS, S. & LEDUC, E. H. (1970). Detection of simultaneous antibody synthesis in plasma cells and specialised lymphocytes in rabbit lymph nodes. *J. exp. Med.* **131,** 1137–68.

AVRAMEAS, S. & LESPINATS, G. (1967). Détection d'anticorps dans des cellules immunocompétentes d'animaux immunisés avec des enzymes. *Compt. Rend. Acad. Sci. (Paris)* **265,** 302–4.

AVRAMEAS, S., TAUDOU, B. & TERNYNCK, T. (1971). Specificity of antibodies synthesised by immunocytes as detected by immunoenzyme techniques. *Int. Arch. Allergy Appl. Immunol.* **40,** 161–70.

AVRAMEAS, S. & TERNYNCK, T. (1969). The cross-linking of proteins with glutaraldehyde and its use for the preparation of immunoadsorbants. *Immunochemistry* **6,** 53–66.

GRAHAM, R. C., Jr. & KARNOVSKY, M. J. (1966). The early stages of absorption of injected horseradish peroxidase in the proximal tubules of mouse kidney; ultrastructural cytochemistry by a new technique. *J. Histochem. Cytochem.* **14,** 291–302.

HAY, J. B., MURPHY, M. J., MORRIS, B. & BESSIS, M. C. (1972). Quantitative studies on the proliferation and differentiation of antibody-forming cells in lymph. *Am. J. Path.* **66,** 1–24.

HESS, M. W., SORDAT, B., STONER, R. D. & COTTIER, H. (1971). Quantitative titration of anti-horseradish peroxidase antibody in mouse serum. *Immunochemistry* **8,** 509–15.

HUGON, J. & BORGERS, M. (1966). A direct lead method for the electron microscopic visualisation of alkaline phosphatase activity. *J. Histochem. Cytochem.* **14,** 429–31.

KRAEHENBUHL, J. P. & CAMPICHE, M. A. (1969). Early stages of intestinal absorption of specific antibodies in the newborn. An ultrastructural, cytochemical, and immunological study in the pig, rat and rabbit. *J. Cell Biol.* **42,** 345–65.

KRAEHENBUHL, J. P., DE GRANDI, P. B. & CAMPICHE, M. A. (1971). Ultrastructural localisation of intracellular antigen using enzyme-labelled antibody fragments. *J. Cell Biol.* **50,** 432–45.

KUHLMANN, W. D. & AVRAMEAS, S. (1971). Glucose oxidase as an antigen marker for light and electron microscopic studies. *J. Histochem. Cytochem.* **19,** 361–8.

KUHLMANN, W. D. & MILLER, H. R. P. (1971). A comparative study of the techniques for ultrastructural localisation of anti-enzyme antibodies. *J. ultrastruct. Res.* **35,** 370–85.

LEDUC, E. H., AVRAMEAS, S. & BOUTEILLE, M. (1968). Ultrastructural localisation of antibody in differentiating plasma cells. *J. exp. Med.* **127,** 109–18.

LEDUC, E. H., SCOTT, G. B. & AVRAMEAS, S. (1969). Ultrastructural localisation of intracellular immune globulins in plasma cells and lymphoblasts by enzyme-labelled antibodies. *J. Histochem. Cytochem.* **17,** 211–24.

MILLER, H. R. P. (1970). A combined light and electron microscopic technique for the localisation of antibody producing cells in the stimulated lymph node. 7th Int. Congr. Electron Microscopy, Vol. 1, 537–8.

MILLER, H. R. P. & AVRAMEAS, S. (1971). Association between macrophages and specific antibody producing cells. *Nature (New Biol.)* **229,** 184–5.

MILLER, H. R. P., AVRAMEAS, S. & TERNYNCK, T. (1972). Differences between antibody-forming cells responding to primary and to secondary stimulation with horseradish peroxidase. *Immunology*, ms. submitted for publication.

SCOTT, G. B. & AVRAMEAS, S. (1968). Intracellular antibody formation demonstrated on ultrathin frozen sections with alkaline phosphatase used as an antigen and as a marker. *Electron Microsc. Proc. 4th Eur. Reg. Conf.* **2,** 201–2.

SCOTT, G. B., AVRAMEAS, S. & BERNHARD, W. (1968). Etude au microscope électronique de la formation d'anticorps à l'aide de phosphatase alcaline utilisée comme antigène. *Compt. Rend. Acad. Sci. (Paris)* **266,** 746–8.

SORDAT, B., SORDAT, M., HESS, M. W., STONER, R. D. & COTTIER, H. (1970). Specific antibody within germinal centre cells of mice after primary immunisation with horseradish peroxidase: a light and electron microscopic study. *J. exp. Med.* **131,** 77–91.

STRAUS, W. (1968). Cytochemical detection of antibody to horseradish peroxidase in spleen and lymph nodes. *J. Histochem. Cytochem.* **16,** 237–48.

STRAUS, W. (1970*a*). Localisation of antibody to horseradish peroxidase in popliteal lymph nodes of rabbits during the primary and early secondary responses. *J. Histochem. Cytochem.* **18,** 120–30.

STRAUS, W. (1970*b*). Localisation of the antigen in popliteal lymph nodes of rabbits during the formation of antibodies to horseradish peroxidase. *J. Histochem. Cytochem.* **18,** 131–42.

Enzyme markers: their linkage with proteins and use in immuno-histochemistry

STRATIS AVRAMEAS*

Laboratoire Chimie des Protéines,
Institut de Recherches Scientifiques
sur le Cancer, Villejuif, France

Contents

Synopsis. The preparation and use of enzyme-labelled antibodies and antigens is described. First, the enzyme is linked covalently to the antibody or antigen. Next, the enzyme-labelled protein is allowed to react with the cellular antigen or antibody; and finally, the sites of bound enzyme are revealed with appropriate cytochemical staining techniques at either the light or electron microscopical levels.

Because the specific activity of an enzyme can be assayed by appropriate enzymological techniques, enzyme-labelled antibodies can also be employed for measuring the amounts of cellular constituents. Similarly, enzyme-labelled antigens can be used for the quantitation of humoral antigens.

In addition to the antigen-antibody reaction, enzyme markers can also be used for the quantitation and localization of other specifically interacting constituents.

---

* Present address: Unité d'Immunocytochimie, Département de Biologie Moléculaire, Institut Pasteur, 28 rue du Dr. Roux, 75-Paris XV°, France.

Introduction

Since immunofluorescence was introduced by Coons & Kaplan (1950) as a technique for the localization of cellular constituents, considerable progress in immunocytological procedures has been achieved. The common feature of all these procedures is a marker substance which is attached to an antibody, visible at the light or electron microscope levels or both. The labelled antibody preparation is then used for the detection of particular cellular constituents. Fluorescent (Coons, 1956) and radioactive (Berenbaum, 1959) substances, ferritin (Singer, 1959) and heavy metals (Pepe, 1961) have been used as the markers.

The detection of cellular constituents at the light microscope level has been carried out mainly with the immunofluorescence technique while the immunoferritin method has been used almost exclusively for ultrastructural studies.

More recently, the use of enzymes as markers was introduced (Nakane & Pierce, 1966; Avrameas & Uriel, 1966). In these procedures, the enzyme is first covalently-linked to the antibody; then the enzyme-labelled antibody is allowed to react with the cellular antigen and finally, the sites of enzyme activity are revealed by appropriate cytochemical methods. Since specific cytochemical techniques for staining enzymes are available both for light and electron microscopy, enzyme-labelled antibodies can be employed equally well for microscopic studies at the two levels.

Methods available

When dealing with enzyme-immunocytochemical methods, one should bear in mind that the choice of the three basic elements, namely antibody, enzyme, and cross-linking reagent, is of major importance.

(a) *Antisera and antibodies*

For the labelling, either the gamma-globulin fraction from immune sera or specifically purified antibodies can be used. The best results in both cases are obtained when the starting material is a hyperimmune serum containing high levels of antibody. Although satisfactory results can be obtained and have been reported (Benson *et al.*, 1970; Fukuyama *et al.*, 1970; Petts & Roitt, 1971) with immune gamma-globulins, it is preferable to employ, whenever possible, pure antibodies. The results obtained with pure labelled-antibodies are much more reproducible and non-specific background staining is much less pronounced.

The isolation of pure antibodies is accomplished by the use of the immunoadsorption techniques, using water insoluble protein derivatives (Silman & Katchalski, 1966). An easy way to prepare them is to employ glutaraldehyde (Avrameas & Ternynck, 1969). The protein to be insolubilized is dissolved in a buffer of near neutral pH. Then glutaraldehyde is added and a gel will form 10 to 30 min later, depending on the protein. After dispersion, homogenization and washing, the polymer is ready for use as an immunoadsorbent for the isolation of antibodies. It is worth while noting that by using glutaraldehyde, all the proteins present in a serum, or in a biological fluid, in general will be insolubilized. These insoluble derivatives can be employed successfully in the preparation of monospecific antisera or the isolation of specific antibodies.

Another simple way of preparing immunoadsorbents is to couple proteins to polyacrylamide beads activated with glutaraldehyde (Ternynck & Avrameas, 1972). By this procedure, small quantities of antigen are sufficient to prepare effective immunoadsorbents.

(b) *Enzyme markers*

Obviously, any enzyme that is detectable by histochemical methods (Pearse, 1960) and that does not modify the structure of the cell can be utilized as a marker. However, the enzyme to be used as the label should possess one, if not all, of the following criteria: first, a high specific activity and turnover number; second, a relatively high stability at room temperature; third, it should not lose a substantial part of its activity after a coupling procedure; and finally, it should be commercially available. This explains why at present the enzymes that can be used as labelling substances are limited to a small number. Amongst them, horseradish peroxidase is the enzyme of first choice. *E. Coli* alkaline phosphatase and *Aspergilus niger* glucose oxidase, although not often employed, have also given reproducible and satisfactory results (Avrameas, 1969; Avrameas *et al.*, 1971). Specific cytochemical techniques for localizing these enzymes are available for both light and electron microscopy (Graham & Karnovsky, 1966; Hugon & Borgers, 1966; Kühlmann & Avrameas, 1971). However, staining for peroxidase as compared to the two other enzymes, permits a more precise localization and identification of ultrastructures, and it is for this reason that peroxidase is almost exclusively employed in electron microscopy.

(c) *Cross-linking agents for the coupling of enzymes to antibodies*

For the coupling of enzymes with antibodies, the following cross-linking agents have been employed: *p,p'*-difluoro-*m,m'*-dinitrophenyl sulphone and water-soluble carbodiimides (Nakane & Pierce, 1967); tetrazotized *o*-dianisidine, cyanuric chloride and glutaraldehyde (Avrameas, 1968). In our hands, the most satisfactory and reproducible results were obtained with glutaraldehyde by employing the following technique: the antibody and the enzyme to be coupled are dissolved in a phosphate buffer of near neutral pH and then glutaraldehyde is added dropwise to the stirred solution. After 3 hr, the solution is dialyzed, centrifuged and stored at 4°C until required for use (Avrameas, 1968, 1969*a*). Using this procedure, several enzymes have been coupled to various antibodies and the conjugates obtained were found to be highly effective for the immunohistochemical detection of the corresponding antigens. The conjugates remained immunologically and enzymatically active, even after two years of storage at 4°C. Experiments performed using several techniques have shown that the enzyme antibody conjugates prepared with glutaraldehyde were composed of a highly heterogeneous population of complexes but that free antibody was absent in the reaction mixture. The average ratio of antibody to the enzyme in the active complexes varied between 1 and 3.

Recently, a two-step procedure was developed which uses glutaraldehyde to prepare both peroxidase-antibody and peroxidase-Fab conjugates (Avrameas & Ternynck, 1971). This procedure is based on the observation that peroxidase alone cannot be insolubilized even with an excess of glutaraldehyde. In fact, peroxidase possess few lysine residues (Welinder *et al.*, 1972), residues with which almost exclusively glutaraldehyde reacts, and probably most of them have the ε-amino group blocked. Thus, when glutaraldehyde is added in excess to a solution of peroxidase, it is conceivable that most of the glutaraldehyde reacts with the free amino groups of peroxidase only by one of its two active aldehyde groups. The remaining free aldehyde group should then be available for combination with the amino groups of a protein added subsequently.

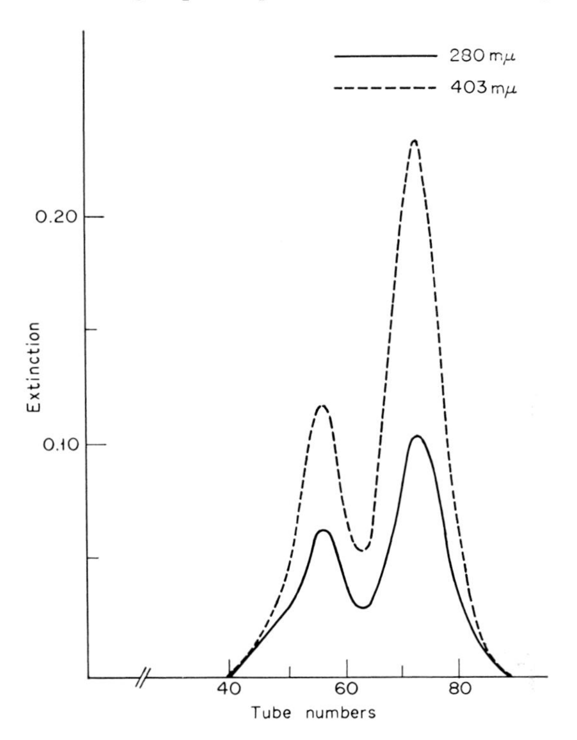

Figure 1. Chromatography on Sephadex G-200 column (100 × 2 cm) equilibrated with 0.1 M phosphate buffer, pH 6.8, and calibrated for mol. wt. determinations of peroxidase-labelled sheep Fab anti-rabbit IgG prepared by the two-step procedure. The first peak corresponds to a mol. wt. of 80 000 and the second to one of 40 000. (——) is extinction at 280 nm and (-----) extinction at 403 nm). Extinction values at 403 nm reveal the chromatographic profile of peroxidase (see Avrameas & Ternynck, 1971).

To couple peroxidase to antibody by this procedure, the peroxidase is first dissolved in a phosphate buffer near neutrality and then an excess of glutaraldehyde is added. The reaction mixture is allowed to stand at room temperature overnight and filtered through a Sephadex G-25 column in order to remove the excess of unreacted glutaraldehyde. The peroxidase preparation in which now active aldehyde groups have been introduced is allowed to react at pH 9.5 with the antibody or its Fab fragments. Ultracentrifugal analysis, chromatography on Sephadex G-200 columns equilibrated for mol. wt. determinations and sodium dodecyl sulphate-acrylamide electrophoresis have shown that, by this procedure, only a homogeneous population of conjugate is obtained in which the molar ratio of antibody or Fab to peroxidase is 1:1 (Avrameas & Ternynck, 1971). It is interesting to note that treatment of peroxidase with glutaraldehyde stabilizes the enzyme considerably. Thus, the glutaraldehyde-treated peroxidase is stable under strong denaturating conditions under which the native peroxidase completely loses its catalytic activity. Furthermore, when peroxidase was coupled with other proteins (albumin, soya bean trypsin inhibitor, insulin) using the above procedure, homogeneous populations of conjugates were always obtained in which the molar ratio of peroxidase to the protein was 1:1.

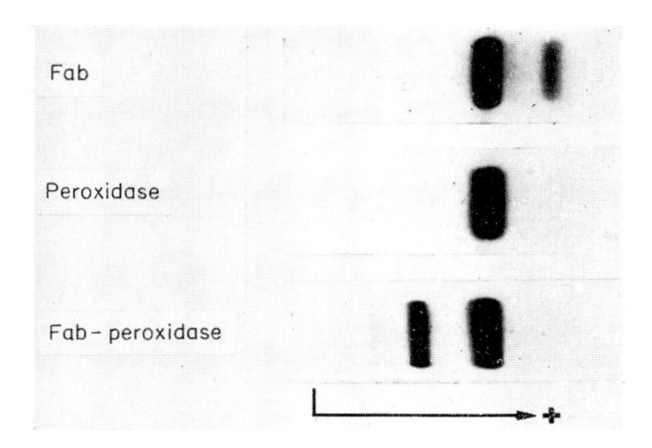

Figure 2. Sodium dodecyl sulphate-acrylamide gel electrophoresis of sheep Fab anti-rabbit IgG; peroxidase; and peroxidase-labelled sheep Fab anti-rabbit IgG prepared by the two-step procedure. After coupling, mainly two electrophoretic bands are seen. The more anodic band corresponds to a mol. wt. of 40 000 and the less anodic one to 80 000 (see Avrameas & Ternynck, 1971).

(d) *Modifications of the enzyme-labelled antibody technique*

In order to avoid coupling procedures or to increase the sensitivity, several variants of the enzyme-labelled antibody technique have been developed (Avrameas, 1969*b*; Mason *et al.*, 1969; Sternberger & Cuculis, 1969). These procedures are based on the principle that enzyme-antienzyme complexes always possess catalytic activity, hence permitting the detection of the enzyme in the immunological complexes (Uriel, 1967). Thus, hybrid antibodies possessing double anti-immunoglobulin antienzyme specificities have been employed for the detection of immunoglobulins and antienzyme antibodies were used to enhance the sensitivity of the immuno-enzyme techniques (Avrameas, 1969*b*). The more characteristic of these procedures is as follows. An anti-immunoglobulin antibody reacts with the immunoglobulin of species A, which is present in a fixed cell in an insoluble form, only by one of its two active sites. Thus, the free active site of the anti-immunoglobulin antibody can link the subsequently added anti-enzyme antibody that was prepared in species A. This anti-enzyme antibody can now bind the finally added enzyme. As the enzyme-antienzyme complex possesses catalytic activity, this allows the cytochemical detection of the enzyme and consequently, the immunoglobulin of species A. Because antibodies are immunoglobulins, it is evident that this procedure allows the detection of any antigen (Avrameas, 1969*b*; Mason *et al.*, 1969; Sternberger & Cuculis, 1969).

Light microscopy

At the light microscope level, the enzyme-labelled antibody technique has been applied with success in various immunocytological studies. Thus, antibodies of various species and specificities were coupled with enzymes and utilized to detect the corresponding antigens (Nakane, 1968, 1970; Baker *et al.*, 1970; Benson *et al.*, 1970; Fukuyama *et al.*, 1970; Avrameas, 1968, 1969*a*; Druhet *et al.*, 1970; Bariety *et al.*, 1971; Zeromski, 1970). Conversely, antigens were coupled with enzymes and employed for detecting antibodies (Avrameas, 1968, 1969*a*; Johnson *et al.*, 1971). In these studies, the localization of antigens or antibodies was carried out on imprints, cell suspensions or tissue sections after adequate fixation. The best results were obtained when dehydrating fixatives like ethanol or acetone were used but good results have also been reported when formaldehyde was used as the fixative (Wicker & Avrameas, 1969). Because for enzymes, histochemical staining techniques giving rise to final reaction products of various colours are available, enzyme-labelled antigen or antibody can be used for the simultaneous detection of two different cellular constituents (Avrameas, 1969*a*; Avrameas *et al.*, 1971; Nakane, 1968; Wicker & Avrameas, 1969). Alternatively, paired staining techniques combining autoradiography or immunofluorescent and immuno-enzymological methods can be employed for the simultaneous or two-step detection of two constituents (Guillien *et al.*, 1968, 1970; Wicker & Avrameas, 1970). Comparison of peroxidase- and fluorescein-conjugated antisera have shown that the sensitivities of both methods were similar and that the localization of the antigen by the two labels were identical (Avrameas, 1969; Benson *et al.*, 1970; Petts & Roitt, 1971; Wicker & Avrameas, 1969).

Electron microscopy

Satisfactory and reproducible results have been obtained when the peroxidase-labelled antibody technique was applied for the ultrastructural localization of cell surface antigen (Bretton & Lespinats, 1969; Bretton *et al.*, 1972; Gonatas *et al.*, 1972). Equally satisfactory results were obtained when the modification of the peroxidase-labelled technique described earlier was used (Willingham *et al.*, 1971).

A systematic comparison between ferritin-labelled and peroxidase-labelled antibody for the localization of cell surface antigens has shown that the peroxidase technique reveals more antigenic determinants than the ferritin one and is several times more sensitive (Bretton *et al.*, 1972).

When the immunoenzymatic procedures were tested for their capacity to reveal intracellular antigen at the ultrastructural level, satisfactory, but very often inconsistant, results were obtained (Kawarai & Nakane, 1970; Leduc *et al.*, 1969*a*; Nakane & Pierce, 1967; Wolff & Schreiner, 1971). This was mainly due to penetration difficulties of the conjugates into the interior of cells fixed by fixatives routinely employed in electron microscopy. In our hands, the most satisfactory results were obtained when Fab-peroxidase conjugates prepared by the two-step procedures described above were used (Kühlman *et al.*, 1972). It must be stressed, however, that although these conjugates are noticeably more effective for the ultrastructural localization of antigens and good results are obtained, difficulties of penetration always persist.

Recently, Fab-antibody fragments labelled either with cytochrome C or peroxidase were compared for their effectiveness for the ultrastructural localization of antigens (Kraehenbuhl *et al.*, 1971). Both markers revealed the antigen equally well but a heavier background was reported to occur with peroxidase. In this connection, it should be stressed once more that for all immunoenzymatic techniques, the most purified peroxidase preparation (RZ*=3) should be used in order to reduce or avoid non-specific staining and it appears more than probable that an impure peroxidase preparation (RZ=0.6) was employed by Kraehenbuhl *et al.* It seems to me that in ultrastructural immunoenzymatic techniques, the limiting factor is rather the shape of the antibody or its Fab fragment rather than the mol. wt. of the label used. In my opinion, further improvements for the ultrastructural localization of antigens should be focused instead on the use of new fixatives and embedding media for electron microscopy.

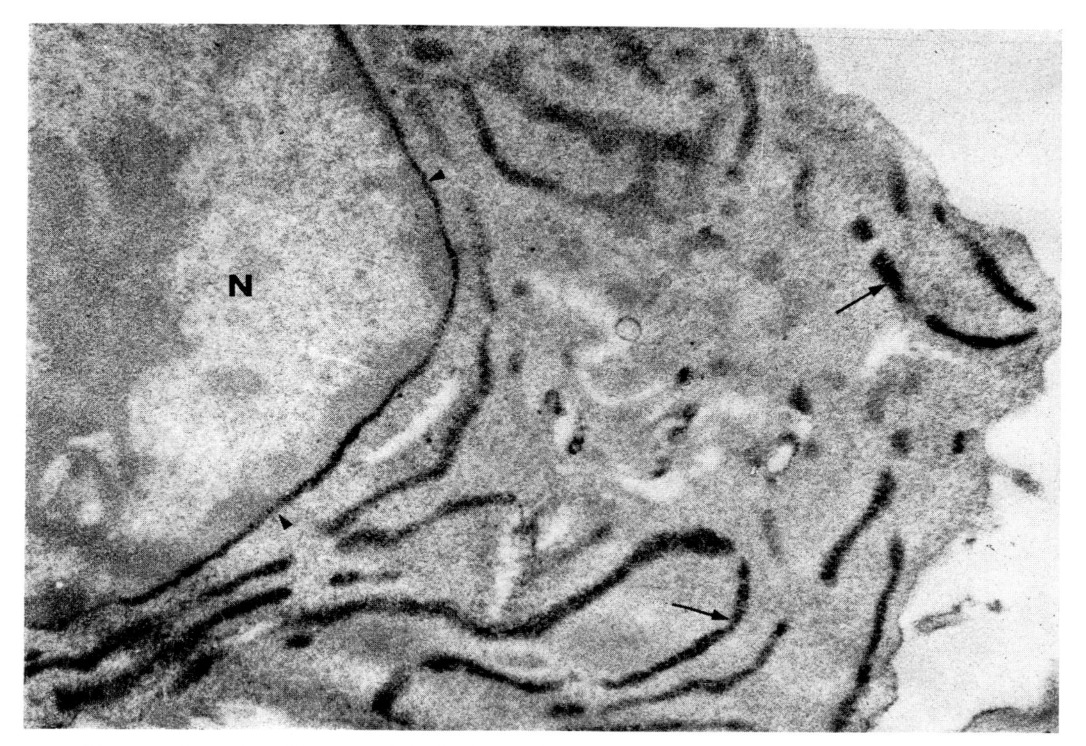

Figure 3. Detection of immunoglobulins in rabbit lymph node by the use of sheep Fab anti-rabbit immunoglobulins labelled with peroxidase. Peroxidase reaction positive in perinuclear space (▲) and in the ergastoplasmic cisternae (arrows); N=nucleus × 18 000 from Kühlman *et al.* (1972), ms. in preparation.

Quantitative techniques with enzyme-labelled proteins

The quantity of an enzyme present in a cell can be accurately determined if use is made of the procedure currently employed in enzymology for the measurement of the catalytic activity of enzymes. It appears then that an enzyme-labelled antibody can be used to measure the activity of cellular constituents. Thus, rabbit antibodies anti-rat IgG labelled with peroxidase have been employed for the dosage of the antigenic determi-

* RZ (*German* Reinheit Zahl) is the ratio of the extinctions at 403 nm (haemin) and 275 nm (protein) of a peroxidase and indicates the degree of purity of peroxidase. For pure crystalline horseradish peroxidase RZ is almost 3.

nants of rat IgG present at the surface of lymphoid cells (Avrameas & Guilbert, 1971*a*). Cell suspensions from lymph nodes were prepared and a given number of cells were incubated for various periods of time at 4°C with increasing concentrations of peroxidase-labelled antibody. The cells were washed by centrifugation and the peroxidase activity of the cellular preparation was determined spectrophotometrically. Knowing the number of the cells, the total quantity of peroxidase fixed by these cells and the ratio of peroxidase to antibody, the mean quantity of antibody fixed per lymphoid cell was calculated. It was thus found that each lymphoid cell carrying IgG possess approximately 230,000 IgG antigenic determinants on its surface.

Recently, peroxidase- and alkaline phosphatase-labelled antigens were also employed for the quantitation of humoral constituents (Avrameas & Guilbert, 1971*b*; Engvall & Perlmann, 1971; Van Weemen & Schuurs, 1971). These procedures are based on the same principles as those developed for quantitative radioimmunoassays. Thus, a fixed quantity of an enzyme-labelled antigen is incubated with increasing concentrations of an unlabelled one in the presence of a given amount of antibody directed against this antigen. The methods seem quite promising since by these assays, 1-100 ng/ml of antigen could be determined.

Conclusions

It appears that the main advantage of the immunoenzyme techniques is the fact that the same reagent, namely the peroxidase-labelled antigen or antibody, can be employed both for quantitative studies and qualitative work at the light and electron microscope levels. In consequence, the results obtained by the different procedures can be directly linked and interpreted objectively.

The use of an enzyme-labelled antibody for the detection of an antigen is based on the general principle that antibodies react specifically with their corresponding antigens. In addition to the antigen-antibody reaction, there are other systems where analogous procedures may be employed. Thus, a procedure using concanavalin A, a saccharide-binding protein, and peroxidase has been developed for the quantitation and the light and electron microscopic localization of cellular carbohydrate components (Avrameas, 1970; Bernhard & Avrameas, 1971). Furthermore, peroxidase-labelled soya bean trypsin inhibitor has been employed for the detection of trypsin in pancreatic cells; and plant agglutinins conjugated with peroxidase have been used for the light and electron microscopic detection of specific cellular carbohydrate components (Gonatas & Avrameas, 1972).

References

AVRAMEAS, S. (1968). Detection d'anticorps et d'antigènes à l'aide d'enzymes. *Bull. Soc. Chim. Biol.* **50,** 1169–78.

AVRAMEAS, S. (1969*a*). Coupling of enzymes to proteins with glutaraldehyde. Use of the conjugates for the detection of antigens and antibodies. *Immunochemistry* **6,** 43–52.

AVRAMEAS, S. (1969*b*). Indirect immunoenzyme techniques for the intracellular detection of antigens. *Immunochemistry* **6,** 825–31.

AVRAMEAS, S. (1970). Emploi de la concanavaline-A pour l'isolement, la détection et la mesure des glycoprotéines et glucides extra- ou endocellulaires. *C.R. Acad. Sci.* (Paris) **270,** 2205–8.

AVRAMEAS, S. & GUILBERT, B. (1971*a*). A method for quantitative determination of cellular immunoglobulins by enzyme-labeled antibodies. *Europ. J. Immunol.* **1,** 394–6.

AVRAMEAS, S. & GUILBERT, B. (1971*b*). Dosage enzymo-immunologique de protéines à l'aide d'immunoadsorbants et d'antigènes marqués aux enzymes. *C.R. Acad. Sci.* (Paris) **273,** 2705–7.

AVRAMEAS, S., TAUDOU, B. & TERNYNCK, T. (1971). Specificity of antibodies synthetized by immunocytes as detected by immunoenzyme techniques. *Int. Arch. Allergy* **40,** 161–70.

AVRAMEAS, S. & TERNYNCK, T. (1969). The cross-linking of proteins with glutaraldehyde and its use for the preparation of immunoadsorbents. *Immunochemistry* **6,** 53–63.

AVRAMEAS, S. & TERNYNCK, T. (1971). Peroxidase labelled antibody and Fab conjugates with enhanced intracellular penetration. *Immunochemistry* **8,** 1175–9.

AVRAMEAS, S. & URIEL, J. (1966). Méthode de marquage d'antigènes et d'anticorps avec des enzymes et son application en immunodiffusion *C R. Acad. Sci.* (Paris) **262,** 2543–5.

BAKER, B. L., PEK, S., MIDGLEY, A. R. & GERSTEN, B. E. (1970). Identification of the corticotropin cell in Rat hypophyses with peroxidase-labelled antibody. *Anat. Rec.* **166,** 557–63.

BARIETY, J., DRUET, P., LALIBERTE, F. & SAPIN, C. (1971). Glomerulonephritis with γ and f_1-C globulin deposits induced in rats by mercuric chloride. *Amer. J. Path.* **65,** 293–302.

BENSON, M. D. & COHEN, A. S. (1970). Antinuclear antibodies ins sytemic lupus erythematosus. *Ann. Int. Med.* **73,** 943.

BERENBAUM, M. D. (1959). The autoradiographic localization of intracellular antibody. *Immunology* **2,** 71–83.

BERNHARD, W. & AVRAMEAS, S. (1971). Ultrastructural visualization of cellular carbohydrate components by means of concanavalin-A. *Exp. Cell Res.* **64,** 232–6.

BRETTON, R. & LESPINATS, G. (1969). Localization ultrastructurale d'antigènes à la surface de cellules tumorales. *C.R. Acad. Sci.* (Paris) **268,** 3223–5.

BRETTON, R., TERNYNCK, T. & AVRAMEAS, S. (1972). Comparison of peroxidase and ferritin labelling of cell surface antigens. *Exp. Cell Res.* **71,** 145–55.

COONS, A. H. (1956). Histochemistry with labeled antibody. *Int. Rev. Cytol.* **5,** 1–23.

COONS, A. H. & KAPLAN, M. H. (1950). Localization of antigen in tissue cells. *J. exp. Med.* **91,**1.

DRUET, P., LELOUP, B., BARIETY, J. & LAGRUE, G. (1970). Application à la pathologie rénale de la technique des anticorps couplés aux enzymes. *Rev. Europ. Etudes Clin. Biol.* **15,** 119–22.

ENGVALL, E. & PERLMANN, P. (1971). Enzyme-linked immunoadsorbent assay (ELISA) Quantitative assay of immunoglobulin G. *Immunochemistry* **8,** 871–4.

FUKUYAMA, K., DOUGLAS, S. D., TUFFANELLI, D. L. & EPSTEIN, W. L. (1970). Immunohistochemical method for localization of antibodies in cutaneous disease. *Am. J. Clin. Path.* **54,** 410–18.

GONATAS, N. K., ANTOINE, J. C., AVRAMEAS, S. & STIEBER, A. (1972). Surface immunoglobulins of thymus and lymph node cells demonstrated by the peroxidase coupling technique. *Lab. Invest.* **26,** 253–61.

GONATAS, N. K. & AVRAMEAS, S. (1972). Cell affinity labelling with peroxidase conjugates (in preparation).

GRAHAM, R. C. & KARNOVSKY, M. J. (1966). The early stages of absorption of injected horseradish peroxidase into the proximal tubules of mouse kidney. Ultrastructural cytochemistry by a new technique. *J. Histochem. Cytochem.* **14,** 291–302.

GUILLIEN, P., AVRAMEAS, S. & BURTIN, P. (1970). Specificity of antibodies in single cells after immunization with antigens bearing several antigenic determinants. *Immunology* **18,** 483–91.

GUILLIEN, P., BURTIN, P. & AVRAMEAS, S. (1968). Association de l'immunofluorescecne et de l'immunoenzymologie pour la détection d'anticorps intra-cellulaires. *C.R. Acad. Sci.* (Paris) **267,** 1425–7.

HUGON, J. & BORGERS, M. (1966). Ultrastructural localization of alkaline phosphatase activity in the absorbing cells of the duodenum of mouse *J. Histochem. Cytochem.* **14,** 629–40.

JOHNSON, A. B., WISNIEWSKI, H. M., RAINE, C. S., EYLAR, E. H. & TEARY, R. D. (1971). Specific binding of peroxidase labeled myelin basic protein in allergic encephalomyelitis. *Proc. Natl Acad. Sci.* **68,** 2694–8.

KAWARAI, Y. & NAKANE, P. K. (1970). Localization of tissue antigens on the ultrathin sections with peroxidase labelled antibody method. *J. Histochem. Cytochem.* **18,** 161–6.

KRAEHENBUHL, J. P., DE GRANDI, P. D. & CAMPICHE, M. A. (1971). Ultrastructural localization of intracellular antigen using enzyme labeled antibody fragments. *J. Cell Biol.* **50,** 432–45.

KÜHLMAN, W. & AVRAMEAS, S. (1971). Glucose-oxidase as an antigen marker for light and electron microscopic studies. *J. Histochem. Cytochem.* **19,** 361–8.

KÜHLMAN, W., AVRAMEAS, S. & TERNYNCK, T. (1972). Comparison of various antibody-peroxidase conjugates for the ultrastructural localization of antigens (*ms. in preparation*).

LEDUC, E. H., SCOTT, G. B. & AVRAMEAS, S. (1969). Ultrastructural localization of intracellular immunoglobulins in plasma cells and lymphoblasts by enzyme labeled antibodies. *J. Histochem. Cytochem.* **17,** 211–24.

LEDUC, E. H., WICKER, R., AVRAMEAS, S. & BERNHARD, W. (1969). Ultrastructural localization of SV 40 T antigen with enzyme labelled antibody. *J. Gen. Virol.* **4,** 609–14.

MASON, T. E., PHIFER, R. F., SPICER, S. S., SWALLOW, R. A. & DRESKIN, R. B. (1969). An immunoglobulin-enzyme bridge method for localizing tissue antigens. *J. Histochem. Cytochem.* **17,** 563–9.

NAKANE, P. K. (1968). Simultaneous localization of multiple tissue antigens using the peroxidase-labeled antibody method. A study on pituitary glands of the Rats. *J. Histochem. Cytochem.* **16,** 557–60.

NAKANE, P. K. (1970). Classifications of anterior puitary cell types with immunoenzyme histochemistry. *J. Histochem. Cytochem.* **18,** 9–20.

NAKANE, P. K. & PIERCE, G. B. (1966). Enzyme labeled antibodies. Preparation and application for the localization of antigens. *J. Histochem. Cytochem.* **14,** 929–31.

NAKANE, P. K. & PIERCE, G. B. (1967). Enzyme-labeled antibodies for the light and electron microscopic localization of tissue antigens. *J. Cell Biol.* **33,** 307–18.

PEARSE, A. G. E. (1960). *Histochemistry: Theoretical and Applied,* 2nd Edn. (Churchill: London).

PEPE, F. A. (1961). The use of specific antibody in electron microscopy. Preparation of mercury-labeled antibody. *J. biophys. biochem. Cytol.* **11,** 515–20.

PETTS, V. & ROITT, I. M. (1971). Peroxidase conjugates for the demonstration of tissue antibodies: evaluation of the technique. *Clin. exp. Immunol.* **9,** 407–18.

SILMAN, I. H. & KATCHALSKI, E. (1966). Water-insoluble derivatives of enzymes, antigens and antibodies. *Ann. Rev. Biochem.* **35,** 873–908.

SINGER, S. J. (1959). Preparation of an electron dense antibody conjugate. *Nature, Lond.* **183,** 1523–4.

STERNBERGER, L. A. & CUCULIS, J. J. (1969). Method for enzymatic intensification of the immunocytochemical reaction without use of labelled antibodies. *J. Histochem. Cytochem.* **17,** 190.

TERNYNCK, T. & AVRAMEAS, S. (1972). Polyacrylamide-protein immunoadsorbents prepared with glutaraldehyde. *FEBS Letters* **23,** 24–8.

URIEL, J. (1967). In: *Antibodies to Biologically Active Molecules* (ed. B. Cinader), pp. 181–96. Macmillan & Pergamon: New York.

VAN WEEMEN, B. K. & SCHUURS, A. H. W. M. (1971). Immunoassay using antigen-enzyme conjugates. *FEBS Letters* **15,** 232–6.

WELINDER, K. G., SMILLIE, L. B. & SCHONBAUM, G. R. (1972). Amino-acid sequence studies of horseradish peroxidase. *Can. J. Biochem.* **50,** 44–90.

WICKER, R. (1971). Comparison of immunofluorescent and immunoenzymatic techniques applied to the study of viral antigens. *Ann. N.Y. Acad. Sci.* (USA) **177,** 490–500.

WICKER, R. & AVRAMEAS, S. (1969). Localization of virus antigens by enzyme-labeled antibodies. *J. Gen. Virol.* **4,** 465–71.

WICKER, R. & AVRAMEAS, S. (1970). Application de l'autoradiographie associée aux techniques immuno-enzymatiques à l'étude des antigènes et des anticorps. *C.R. Acad. Aci.* (Paris) **270,** 431–3.

WILLINGHAM, M. C., SPICER, S. S. & GRABER, D. C. (1971). Immunocytologic labeling of calf and human lymphocyte surface antigens. *Lab. Invest.* **25,** 211–19.

WOLFF, K. & SCHREINER (1971). Ultrastructural localization of pemphigus antibodies within the epidermis. *Nature, Lond.* **229,** 59–61.

ZEROMSKI, J. (1970). Immunological findings in sensory carcinomatous neuropathy. Application of peroxidase labelled antibody. *Clin. exp. Immunol.* **6,** 633–7.

Fixation: What should the pathologist do?

I. M. P. DAWSON*

Department of Clinical Histochemistry,
Westminster Medical School,
Udall Street Laboratories, London

Synopsis. Pathologists handling surgical material must make a definitive diagnosis, usually on a Haematoxylin-Eosin stained section, but must also consider beforehand what supporting histochemical investigations may be needed. When only one small piece of tissue is available it should either be fixed in cold neutral or buffered formalin-calcium or formalin-sucrose or alternatively quenched; the first alternative offers a more practical all-round procedure and most investigations which are likely to be of use can be done following it. If the biopsy is divisible into two, one piece should be fixed and the second quenched; if 3 pieces are available, the third should be fixed in glutaraldehyde for possible electron microscopy.

Introduction

The practising pathologist's first duty is to his patient. It is his task to examine carefully, in the most appropriate way, any tissue which the surgeon may remove, make as accurate a diagnosis as he can of the nature and meaning of any changes in it and try to assess prognosis (Dawson, 1972). The way he sets about this must be primarily determined by the provisional clinical diagnosis already made and by the nature of the tissue available to him. There is often sufficient material to allow a number of representative tissue blocks to be processed, using selected methods of preservation, each one suitable for a particular histochemical technique. Sometimes, however, a small piece of tissue only is removed as a biopsy for the sole purpose of diagnosis and thus only one block, or at the most two blocks, is or are available for study. This paper is concerned with the preservation of surgical material in order to obtain the maximum possible information.

The first essential is for the clinician and pathologist to discuss the possible illnesses from which the patient may suffer and which they expect the biopsy to confirm. They should do this before any biopsy is taken. Most histopathologists will derive the maximum information from a thin (5 μm) section of tissue conventionally stained with Haematoxylin or Haemalum and Eosin. On occasion however, other techniques may be more valuable; in suspected primary disaccharidase deficiency for example, when a small intestinal biopsy is taken, conventional histology is usually normal, and it may be pre-

* Now: Department of Pathology, University of Nottingham.

ferable to homogenise the biopsy and perform quantitive assays of disaccharidases (Nordström *et al.*, 1967, 1969), unless the histochemical indigogenic substrates described by Lojda (1970*a,b*) are available.

When on the basis of a tentative clinical diagnosis a conventionally stained section is considered necessary, as it normally is, one must consider the possible confirmatory histochemical investigations which may be needed before deciding on conventional processing; for example, on a biospy thought likely to be from a 'carcinoid' tumour, one would seriously consider processing by freeze-drying and formaldehyde vapour fixation if the equipment were available, because it allows a fuller range of histochemical procedures on the granules (Dawson, 1970), while a liver biopsy from a suspected glycogen storage disorder might be quenched rather than conventionally fixed (Lake, 1970) to prevent glycogen diffusion. The largest part of poor histochemistry results from lack of preplanning or proper application of techniques in the earliest stages of processing and mistakes made here often cannot afterwards be retrieved.

When a provisional diagnosis cannot be made, or when the pattern of possible subsequent histochemical investigations cannot be envisaged, other secondary considerations come into force. The technique of tissue preservation chosen must allow the cutting of good quality sections which can be conventionally stained. It must permit as wide as possible a range of subsequent histochemical procedures once a tentative diagnosis has been made on the stained section. It must allow, where possible, the preservation of material over a number of months or years, and finally it must take account of the laboratory facilities for performing specialized techniques and for storing material. Under these circumstances, two different patterns of tissue preservation must be considered and weighed against one another.

Preservation by fixation

Up to the time of this symposium, one of the best general reviews on fixation was that of Hopwood (1969). The advantages of fixation with aldehydes (Sabatini *et al.*, 1963) are that proteins are bound together by cross-links and bridges, autolysis is prevented, subsequent procedures can be done at leisure, and mucosubstances, most lipids and some enzyme systems are preserved. Disadvantages are that macromolecules are often lost or displaced by diffusion, many enzymes and antigen or antibody-combining sites are affected by the fixative and one does not know precisely after washing which reactive side chains are freed and which remain bound.

Of the various aldehyde fixatives available (Sabatini *et al.*, 1963; Janigan 1964, 1965), formaldehyde is by far the most widely used. It is cheap, and preserves lipids, mucosubstances and hydrolytic enzymes at least as well as other aldehydes. It has the disadvantage of being a less good fixative for subsequent electron microscopy than glutaraldehyde, but recent studies suggest that it preserves tissues better than had been originally thought (Baker & McCrae, 1966; Bradbury & Stoward, 1967). It should always be neutralized and preferably buffered to a slightly alkaline pH. Diffusion of macromolecules can be slowed down by fixation in the cold and by adding calcium to the fixative, and osmotic effects can be minimized by adding sucrose. Material so processed can subsequently either be conventionally embedded in paraffin wax or carbowaxes and sectioned, or frozen and cut directly on the cryostat without embedding. The latter procedure allows techniques for localizing lipids, mucosubstances and many hydrolytic enzymes to be carried out, while still yielding first quality sections for conventional Haematoxylin and Eosin staining.

Glutaraldehyde, a dialdehyde, preserves cytological ultrastructural detail better, but enzyme activities less well; it is a more rapid fixative than formaldehyde, at any rate for albumin (Flitney, 1966), but is more expensive, penetrates tissue blocks less readily and cannot be obtained in other than as a 25% solution. In my view, as in that of others (Chambers *et al.*, 1968) it can replace formaldehyde under some circumstances but does limit the range of hydrolytic enzymes which can subsequently be demonstrated, and does not penetrate well so that the central part of even moderate size blocks may be poorly preserved.

In my opinion, fixation in cold (4°C) neutral or buffered formalin-calcium (Baker, 1946) or buffered formalin-sucrose (Holt *et al.*, 1960) followed by impregnation in gum sucrose, freezing and cryostat sectioning without embedding remains a technique of choice for the small biopsy.

Preservation by freezing

Tissues quenched at low temperatures (below −70°C) very rapidly are well preserved at that temperature; in particular there is no appreciable loss of enzyme activity, diffusion of small molecules is prevented and all potentially reactive chemical groupings remain free. The frozen blocks can be brought to cryostat temperatures (between −20°C and −30°C) for sectioning; where fixation is likely to be advantageous, the fresh cryostat sections can be post-fixed in cold formalin-calcium or other fixative. Contrary to what is often stated, very adequate Haemalum and Eosin staining is possible on fresh frozen, post-fixed sections; they are certainly good enough to allow, for example, a search for ganglion cells in Hirschprung's disease.

Success in freezing techniques, and the avoidance of gross ice-crystal artefacts depends largely on the rapidity of freezing. Techniques which freeze slowly are useless. There is no doubt that the best way of quenching is to use an inert liquid such as Arcton or iso-pentane precooled in liquid nitrogen; specimens are sometimes cooled directly in liquid nitrogen, but this can cause vapourization of the nitrogen immediately surrounding them, leading to an insulating jacket of nitrogen vapour which slows down further cooling. Not all departments, however, have access to liquid nitrogen; isopentane pre-cooled in a dry-ice acetone mixture at −70°C is often a reasonable substitute. The principal disadvantage of quenching is the small size of block which can be handled; in my view, the maximum size is 0.5 × 0.5 × 0.3 cm.

Choice of technique

If the specimen is large, I prefer to take a number of blocks and process as follows:

(*a*) One or two blocks of size about 1.5 × 1.5 × 0.3 cm into neutral or buffered 10% formalin-calcium at 4°C for 18–36 hr, followed by paraffin embedding. These can be used for conventional Haematoxylin & Eosin stains, 'special' histological stains, mucosubstances, 5-hydroxytryptamine fluorescence etc.

(*b*) One or two blocks of similar size into buffered formalin-sucrose at 4°C for 4–18 hr.

depending on the block size, followed by impregnation in gum sucrose for 24 hr or more. These can be used for most hydrolytic enzymes as well as for conventional Haematoxylin & Eosin staining, and once tissue is in gum sucrose it can be stored for a reasonable time.

(*c*) One or two blocks of similar size into buffered or neutral formalin-calcium at 4°C. These can be used for lipids and are, I think, preferable to sucrose-treated material for this purpose.

(*d*) One or two small blocks, up to 0.5 × 0.5 × 0.3 cm, quenched in isopentane pre-cooled in liquid nitrogen, (or in dry ice-acetone) and stored subsequently for short periods at −70°C or in a cryostat cabinet at −30°C. These can be used for dehydrogenases, diaphorases, glycogen and other techniques which are inhibited by fixation or need special and otherwise undesirable fixatives, and cryostat sections from them, if post-fixed or celloidinized, are suitable for all techniques in which fixation is necessary or desirable. The only real disadvantage of this method of preservation is the small size of the blocks which can be frozen.

(*e*) Three or four pieces (0.1 × 0.1 × 0.1 cm) into buffered glutaraldehyde in case electron microscopy is subsequently needed.

(*f*) One or more blocks preserved suitably for any special techniques which may be indicated.

References to freeze-drying of material, with or without subsequent vapour fixation, are deliberately excluded because facilities are not generally available.

When the specimen is small and has been taken specifically for diagnosis, good conventional histology is of prime importance. Should there be sufficient material for one block only, and provided that no single special investigation is clearly indicated by the presumptive clinical diagnosis, the choice in my view lies between fixation in cold formalin-calcium or formalin-sucrose and quenching. Quenching allows the widest range of histochemical procedures, and where a biopsy is too small to be divided in any event, the limitation of size for frozen material is no problem. It does not allow subsequent electron microscopy, and it does mean that the pathologist or a skilled technician must handle the specimen at the moment of resection; quenching cannot be left to the surgeon. Fixation using some form of formaldehyde restricts the range of subsequent histochemistry, allows some subsequent electron microscopy though under less than optimal conditions, but also allows the surgeon to preserve his own material under reasonable conditions, when biopsies are taken outside normal laboratory hours or a skilled technician is not available. For these reasons I believe that if only one small piece of tissue is available it should be fixed; if the biopsy is divisible into 2, one piece should be fixed and the second quenched, and if divisible into 3 the third should be minced and fixed in buffered glutaraldehyde for possible electron microscopic examination.

Once the biopsy is fixed, the decision has to be made whether or not to embed. The necessary procedures further limit histochemistry, particularly by destroying enzyme activity and extracting lipids, and in the vast majority of instances my own preference is for cryostat sections; those stained conventionally can subsequently be dehydrated and mounted. I would make an exception, possibly, for lymph node biopsies in which nuclear detail can be of great importance, but these are fortunately usually divisible so that one block is available for paraffin embedding.

It is true that fixation is a dangerous process, but the practising pathologist must sometimes live dangerously. As Pearse (1968) has recently stated 'After recent demonstrations of the excellent preservative qualities of formalin in respect of many common enzymes, this fixative is gradually returning to favour with histochemists'.

Acknowledgements

I am grateful to Dr M. I. Filipe and the technical staff of the Udall Street Laboratories for their ready co-operation in dealing with the histochemical problems which arise from a routine pathological practice, and to my surgical and medical colleagues for the patience they have shown in handling biopsies in ways sometimes unfamiliar to them. My thanks are due to Mrs Dora Litinakis who has prepared the manuscript.

References

BAKER, J. R. (1946). The histochemical recognition of lipines. *Quart J. Microscop. Sci.* **87,** 441–70.

BAKER, J. R. & MCCRAE, J. M. (1966). The fine structure resulting from fixation by formaldehyde. The effects of concentration, duration and temperature. *Jl R. microsc. Soc.* **85,** 391–9.

BRADBURY, S. & STOWARD, P. J. (1967). The specific cytochemical demonstration in the electron microscope of periodate-reactive mucosubstances and polysaccharides containing *vic*-glycol groups. *Histochemie* **11,** 71–80.

CHAMBERS, R. W. BOWLING, M. C. & GRIMLEY, P. M. (1968). Glutaraldehyde fixation in routine pathology. *Arch. Path.* **85,** 18–30.

DAWSON, I. (1970). The endocrine cells of the gastrointestinal tract. *Histochem. J.* **2,** 527–49.

DAWSON, I. M. P. (1972). The histochemistry of Crohn's disease. In *Clinics in Gastroenterology* (ed. B. Brooke), Vol. 1, No. 2, pp. 309–20. London: Saunders.

FLITNEY, F. W. (1966). The time course of the fixation of albumin by formaldehyde, glutaraldehyde, acrolein and other higher aldehydes. *J. R. microsc. Soc.* **85,** 353–64.

HOLT, S. J. HOBBIGER, E. L. & PAWAN, G. L. S. (1960). Preservation of integrity of rat tissues for cytochemical staining purposes. *J. biophys. biochem. Cytol.* **7,** 383–6.

HOPWOOD, D. (1969). Fixatives and fixation: a review. *Histochem. J.* **1,** 323–60.

JANIGAN, D. T. (1964). Tissue enzyme fixation studies I. The effects of aldehyde fixation on β-glucuronidase, β-galactosidase, N-acetyl-β-glucosaminidase and β-glucosidase in tissue blocks. *Lab. Invest.* **13,** 1038–50.

JANIGAN, D. T. (1965). The effects of aldehyde fixation on acid phosphatase activity in tissue blocks. *J. Histochem. Cytochem.* **13,** 476–83.

LAKE, B. D. (1970). The histochemical evaluation of the glycogen storage disorders. A review of techniques and their limitations. *Histochem. J.* **2,** 441–50.

LOJDA, Z. (1970 *a*). Indigogenic methods for glycosidases I. An improved method for β-D glucosidase and its application to localization studies of intestinal and renal enzymes. *Histochemie* **22,** 347–61.

LOJDA, Z. (1970 *b*). Indigogenic methods for glycosidases II. An improved method for β-D galactosidase and its application to localization studies of the enzymes in the intestine and in other tissues. *Histochemie* **23,** 266–88.

NORDSTRÖM, C., DAHLQVIST, A. & JOSEFSSON, L. (1967). Quantitative determination of enzymes in different parts of the villi and crypts of rat small intestine. Comparison of alkaline phosphatase, disaccharidases and dipeptidases. *J. Histochem. Cytochem.* **15,** 713–21.

NORDSTRÖM, C. KOLDOVSKY, O. & DAHLQVIST, A. (1969). Localization of β-galactosidases and acid phosphatase in the small intestinal wall. Comparison of adult and suckling rat. *J. Histochem. Cytochem.* **17,** 341–7.

PEARSE, A. G. E. (1968). *Histochemistry: Theoretical and Applied,* 3rd Edn. Vol. I, p. 601. London: Churchill.

SABATINI, D. D., BENSCH, K., BARRNETT, R. J. (1963). Cytochemistry and electron microscopy. The preservation of cellular ultrastructure and enzymatic activity by aldehyde fixation. *J. Cell Biol.* **17,** 19–58.

Index

PRINTED IN GREAT BRITAIN BY THE BROADWATER PRESS LTD, WELWYN GARDEN CITY, HERTFORDSHIRE